Contents

1 Strict Inequality   **1**
    1.1   Positive Real Number   1
    1.2   Basic Properties of Greater Than   3
    1.3   Further Properties of Greater Than   4
    1.4   Division of Two Inequalities   7
    1.5   Negative Real Number   9
    1.6   Basic Properties of Less Than   10
    1.7   Greater Than and Less Than   13

2 Non-Strict Inequality   **15**
    2.1   Nonnegative Real Number   15
    2.2   Basic Properties of Greater Than or Equal To   16
    2.3   Zero or Negative Real Number   21
    2.4   Basic Properties of Less Than or Equal To   21
    2.5   Definition of Terms   26
    2.6   Integer inequality   28
    2.7   Bernoulli Inequality   33

3 Chained Inequality   **37**
    3.1   Introduction   37
    3.2   Strict Chained Inequality   38
    3.3   Non-Strict Chained Inequality   45
    3.4   Mixed Chained Inequalities   52
    3.5   Extended Chained Inequality   57
    3.6   An Inequality Due to Archimedes   57

4 Absolute Value   **61**
    4.1   Absolute Value   61
    4.2   Absolute Value and Squares   64
    4.3   Absolute Value and Powers   66
    4.4   Absolute Value and Triangle Inequality   69

4.5 Absolute Value and Composite Numbers 76
4.6 Absolute Value and Chained Inequality 78

5 Max and Min Operations 83
5.1 Max Operation 83
5.2 Min Operation 87
5.3 Max and Min Operations and Absolute Value 90

6 Floor and Ceiling 95
6.1 Floor 95
6.2 Basic Properties of the Floor Function 97
6.3 Ceiling 105
6.4 Basic Properties of the Ceiling function 106
6.5 Relationship between the floor and ceiling 113
6.6 Half of an Integer 115

7 Roots 119
7.1 Square Root 119
7.2 Properties of the Square Root 122
7.3 Cube Root 124
7.4 Properties of the Cube Root 126
7.5 Nth Root 128
7.6 Properties of the Nth Root 130

8 Rational Exponents 135
8.1 Power of Power Law 135
8.2 Different Bases Law 140
8.3 Same Base Law 142

9 Logarithms 147
9.1 Introduction to Logarithms 147
9.2 Properties of Logarithms 149
9.3 Logarithms and Exponents 152
9.4 Product and Quotient 154
9.5 Change of Base 156

CHAPTER 1
Strict Inequality

1.1 Positive Real Number

1. In this section, we define positive real number. Then we discuss the closure of the positive real numbers under addition and multiplication. We define the greater than relation and its notation. Then we describe the comma notation and define not greater than.

2. **Definition of positive real number.** A positive real number is a real number greater than zero. It is a number which is the ratio of two positive integers or can be approximated by such a ratio. Examples of positive real numbers are $1, 2, 3, 3 + 1/2$.

3. If x is a positive real number we say x is *positive*. If we say x is positive then x is a positive real number.

4. **Closure of positive real numbers.** Here, we discuss the closure of the positive real numbers under addition and multiplication.

5. The sum of every pair of positive real numbers is positive. For example, since 5 and 6 are positive real numbers, $5 + 6$ must be positive.
    (Positive Real Numbers are Closed Under Addition)

6. The product of every pair of positive real numbers is positive. For example, since 5 and 6 are positive real numbers, $5 \cdot 6$ must be positive.
    (Positive Real Numbers are Closed Under Multiplication)

7. **Definition of greater than.** Consider the real numbers 5 and 4. Observe that $5 - 4 = 1$ is a positive real number. Thus we say 5 is greater than 4.

8. For all real numbers x, y, if $x - y$ is a positive real number then we say x is greater than y. Conversely, if we say x is greater than y then $x - y$ is a positive real number.

(Definition of Greater Than)

9 **Notation of greater than.** The real number 5 is greater than the real number 4. We write this as $5 > 4$. The real number $x+1$ is greater than the real number x. We write this as $x + 1 > x$.

10 Let x, y be real numbers. If x is greater than y, we denote this by writing $x > y$. If we write $x > y$ then x is greater than y.

(Definition of Greater Than Notation)

11 The statement $2^2 = 2 + 2$ is called an equation. The statement $3 > 2$ is called an inequality. For all real numbers x, y, we shall call the statement $x > y$ an inequality.

12 **Comma notation.** We know that 5 and 6 are each greater than 4. Hence $5 > 4$ and $6 > 4$. We write this as $5, 6 > 4$. If x, y, z are real numbers such that $x > z$ and $y > z$ then we write $x, y > z$. If we write $x, y > z$ then $x > z$ and $y > z$.

13 We write $x, y > 0$ if and only if $x > 0$ and $y > 0$. We write $x, y > w$ if and only if $x > w$ and $y > w$.

14 We write $x, y, z > 0$ if and only if $x > 0$ and $y > 0$ and $z > 0$. We write $x, y, z > w$ if and only if $x > w$ and $y > w$ and $z > w$.

15 **Not greater than.** If x is not greater than y, we denote this by writing $x \not> y$. If we write $x \not> y$ then x is not greater than y.

1.2 Basic Properties of Greater Than

1   In this section, we prove some properties of the greater than relation:

name	property
transitivity of inequality	If $x > y$ and $y > z$ then $x > z$.
addition of a constant	If $x > y$ then $x + w > y + w$.
addition of two inequalities	If $x > y$ and $w > z$ then $x + w > y + z$.
multiplication of an inequality by a constant	If $w > 0$ and $x > y$ and $wx > wy$.
multiplication of two inequalities	If $y, z > 0$, $x > y$ and $w > z$ then $xw > yz$.
squaring an inequality	If $y > 0$, $x > y$ then $x^2 > y^2$.

2   Let w, x, y, z be real numbers. Let n be a positive integer.

3   If $x > y$ and $y > z$ then $x > z$.

(Transitivity of Inequality)

Proof   We are given that $x > y$ and $y > z$. Thus $x - y$ and $y - z$ are positive real numbers. Hence $(x - y) + (y - z)$ is positive. Since, $(x - y) + (y - z) = x - z$, it follows that $x - z$ is positive. Therefore $x > z$.   ∎

4   If $x > y$ then $x + w > y + w$.

(Addition of a Constant)

Proof   We are given that $x > y$. Thus $x - y$ is a positive real number. Since $x - y = (x + w) - (y + w)$, it follows that $(x + w) - (y + w)$ is a positive real number. Hence $x + w > y + w$.   ∎

5   If $x > y$ and $w > z$ then $x + w > y + z$.

(Addition of Two Inequalities)

Proof   We are given that $x > y$ and $w > z$. Adding w to both sides of the inequality $x > y$ we obtain $x + w > y + w$. Adding y to both sides of the inequality $w > z$ we obtain $y + w > y + z$. Applying transitivity to the inequalities $x + w > y + w$ and $y + w > y + z$ we get $x + w > y + z$.   ∎

6   If $w > 0$ and $x > y$ then $xw > yw$

(Multiplication of Inequality by a Positive Constant)

Proof We are given that $w > 0$ and $x > y$. Thus w and $x - y$ are positive real numbers. Thus $w(x - y)$ is a positive real number. Hence $xw - yw$ is a positive real number. Therefore $xw > yw$. ∎

7   If $y, z > 0$, $x > y$ and $w > z$ then $xw > yz$.

(Multiplication of Two Inequalities)

Proof By applying transitivity to the inequalities $w > z$ and $z > 0$ we obtain $w > 0$. Hence w and y are positive and we can multiply inequalities by them. Multiplying the inequality $x > y$ by w we obtain $xw > yw$. Multiplying the inequality $w > z$ by y we obtain $wy > zy$. By applying transitivity to the inequalities $xw > yw$ and $yw > yz$ we get $xw > yz$. ∎

8   If $y > 0$ and $x > y$ then $x^2 > y^2$.

Proof

Consider the inequalities $x > y$ and $x > y$. Multiplying both inequalities together, we get $xx > yy$. Therefore $x^2 > y^2$. ∎

9   **Trichotomy.** Trichotomy states that $x > 0$ or $0 > x$ or $x = 0$ for each real number x. We use this axiom to prove the theorem below.

10 $x > y$ or $y > x$ or $x = y$.

Proof Consider the real number $x - y$. Due to the law of trichotomy, $x - y > 0$ or $0 > x - y$ or $x - y = 0$. Add y to both sides of each inequality/equality. Hence $x > y$ or $y > x$ or $x = y$. ∎

1.3 Further Properties of Greater Than

1   In this section, we prove some properties of the greater than relation:

name	property
subtraction of a constant	If $x > y$ then $x - w > y - w$.
negation of an inequality	If $x > y$ then $-y > -x$.
subtraction of two inequalities	If $x > y$ and $w > z$ then $x - w > y - z$.
addition of arbitrary number of inequalities	If $x_1 > y_1, \ldots, x_n > y_n$ then $x_1 + x_2 + \cdots + x_n > y_1 + y_2 + \cdots + y_n$.
multiplication of arbitrary number of inequalities	If $x_1 > y_1, \ldots, x_n > y_n$ then $x_1 \cdot x_2 \cdots x_n > y_1 \cdot y_2 \cdots y_n$.

2 Let w, x, y, z be real numbers. Let n be a positive integer.

3 If $x > y$ then $x - w > y - w$.

(Subtraction of a Constant)

Proof

Since $-w$ is a real number and $x > y$, it follows that $x + (-w) > y + (-w)$. Thus $x - w > y - w$. ∎

4 We know that $6 > 5$. Now compare -5 and -6. Which is greater? Observe that $(-5) - (-6) = -5 + 6$. Thus $(-5) - (-6) = 1$. Hence $(-5) - (-6)$ is indeed a positive real number. Therefore $-5 > -6$.

5 If $x > y$ then $-y > -x$.

(Negation of an Inequality)

Proof We are given that $x > y$. Subtracting x from both sides we obtain $x - x > y - x$. Hence $0 > y - x$. Subtracting y from both sides we get $-y > -y + y - x$. Thus $-y > -x$. ∎

6 If $x > y$ and $w > z$ then $x - z > y - w$.

(Subtraction of Two Inequalities)

Proof We are given that $x > y$ and $w > z$. Negating the inequality $w > z$, we obtain $-z > -w$. Add together the inequalities $x > y$ and $-z > -w$. Hence $x - z > y - w$. ∎

7   If
$$x_1 > y_1, \quad x_2 > y_2, \quad x_3 > y_3$$
then
$$x_1 + x_2 + x_3 > y_1 + y_2 + y_3$$

(Addition of Three Inequalities)

Proof If we add the inequalities $x_1 > y_1$ and $x_2 > y_2$ we get $x_1 + x_2 > y_1 + y_2$. If we add the inequalities $x_1 + x_2 > y_1 + y_2$ and $x_3 > y_3$ we get $x_1 + x_2 + x_3 > y_1 + y_2 + y_3$. ∎

8   If
$$x_1 > y_1, x_2 > y_2, \ldots, x_n > y_n$$
then
$$x_1 + x_2 + \cdots + x_n > y_1 + y_2 + \cdots + y_n$$

(Addition of Arbitrary Number of Inequalities)

Proof If we add the inequalities $x_1 > y_1$ and $x_2 > y_2$ we get $x_1 + x_2 > y_1 + y_2$. If we add the inequalities $x_1 + x_2 > y_1 + y_2$ and $x_3 > y_3$ we get $x_1 + x_2 + x_3 > y_1 + y_2 + y_3$. By continuing in this manner, at some point we would get
$$x_1 + x_2 + \cdots + x_n > y_1 + y_2 + \cdots + y_n$$
∎

9   If $y_1, y_2, y_3 > 0$ and
$$x_1 > y_1, \quad x_2 > y_2, \quad x_3 > y_3$$
then
$$x_1 x_2 x_3 > y_1 y_2 y_3$$

(Multiplication of Three Inequalities)

Proof If we multiply the inequalities $x_1 > y_1$ and $x_2 > y_2$ we get $x_1 x_2 > y_1 y_2$. If we multiply the inequalities $x_1 x_2 > y_1 y_2$ and $x_3 > y_3$ we get $x_1 x_2 x_3 > y_1 y_2 y_3$. ∎

10   Let $y_1, \ldots, y_n > 0$. If
$$x_1 > y_1, \ x_2 > y_2, \ldots, x_n > y_n$$
then
$$x_1 x_2 \cdots x_n > y_1 y_2 \cdots y_n$$

(Multiplication of Arbitrary Number of Inequalities)

Proof If we multiply the inequalities $x_1 > y_1$ and $x_2 > y_2$ we get $x_1 x_2 > y_1 y_2$. If we multiply the inequalities $x_1 x_2 > y_1 y_2$ and $x_3 > y_3$ we get $x_1 x_2 x_3 > y_1 y_2 y_3$. By continuing in this manner, at some point we would get
$$x_1 x_2 \cdots x_n > y_1 y_2 \cdots y_n$$
∎

11   If $y > 0$ and $x > y$ then $x^n > y^n$.

(Raising Both Sides of an Inequality to a Power)

Proof

Consider the n inequalities $x > y$, $x > y$, ..., $x > y$. Multiplying all the inequalities together, we get $x^n > y^n$. ∎

1.4   Division of Two Inequalities

1   In this section, we show that we can divide two inequalities: for all real numbers w, x, y, z, if $w, z > 0$, $x > y$ and $w > z$ then $\frac{x}{z} > \frac{y}{w}$. Before proving this theorem, we prove some lemmas and corollaries.

2   Let w, x, y, z be real numbers.

3   If $x > 0$ then $\frac{1}{x} \neq 0$

Proof Since $x > 0$, it follows that $1/x$ is a real number. Now assume $1/x = 0$. Multiplying both sides of the equality by x we obtain $x(1/x) = x \cdot 0$. Thus $1 = 0$. This is absurd. Hence the assumption that $1/x = 0$ must be wrong. Therefore $1/x \neq 0$. ∎

4 If $x > 0$ then $0 \not> \dfrac{1}{x}$

Proof Since $x > 0$, it follows that $1/x$ is a real number. Assume $0 > 1/x$. Multiplying both sides of the inequality by x we obtain $x \cdot 0 > x(1/x)$. Thus $0 > 1$. This is absurd. Hence the assumption that $0 > 1/x$ must be wrong. Therefore $0 \not> 1/x$. ∎

5 If $x > 0$ then $\dfrac{1}{x} > 0$

Proof Due to trichotomy, $0 > 1/x$ or $1/x = 0$ or $1/x > 0$. The first two cases have been eliminated by previous results. The only remaining case is $\dfrac{1}{x} > 0$. ∎

6 If $w > 0$ and $x > y$ then $\dfrac{x}{w} > \dfrac{y}{w}$

(Division of Inequality by a Constant)

Proof We are given that $w > 0$. Thus $1/w > 0$. Hence we can multiply both sides of the inequality $x > y$ by $1/w$ to obtain $x(1/w) > y(1/w)$. Therefore $x/w > y/w$. ∎

7 If $y > 0$ and $x > y$ then $\dfrac{1}{y} > \dfrac{1}{x}$

(Reciprocal of Inequality)

Proof We are given that $y > 0$. Hence we can divide both sides of the inequality $x > y$ by y to obtain $x/y > 1$.

Applying transitivity to the inequalities $x > y$ and $y > 0$, we get $x > 0$. Thus we can divide both sides of the inequality $x/y > 1$ by x to obtain $1/y > 1/x$. ∎

8 If $y, z > 0$, $x > y$ and $w > z$ then $\dfrac{x}{z} > \dfrac{y}{w}$.

(Division of Two Inequalities)

Proof Since $w > z$, it follows that $1/z > 1/w$. Applying transitivity to the inequalities $w > z$ and $z > 0$ we obtain $w > 0$. Hence $1/w > 0$. Since $y, 1/w > 0$ we can multiply the inequalities $x > y$ and $1/z > 1/w$. We obtain $x(1/z) > y(1/w)$. Therefore $x/z > y/w$. ∎

1.5 Negative Real Number

1. In this section, we define negative real numbers, the less than relation, and its notation.

2. **Definition of negative real number.** Examples of negative real numbers are $-1, -2, -3, -3 + 1/2$. We know that -5 is a negative real number. Observe that $-(-5) = 5$. Hence $-(-5)$ is a positive real number.

3. A real number x is a negative real number if and only if $-x$ is positive.
   (Definition of Negative Real Number)

4. If x is a negative real number we say x is *negative*. If we say x is negative then x is a negative real number.

5. Every real number is positive or negative or zero.

6. **Definition of less than.** Consider the real numbers 5 and 6. Observe that $5 - 6 = -1$ is a negative real number. Thus we say 5 is less than 6.

7. For all real numbers x, y, if $x - y$ is a negative real number then we say x is less than y. Conversely, if we say x is less than y then $x - y$ is a negative real number.
   (Definition of Less Than)

8. **Notation of less than.** The real number 5 is less than the real number 6. Hence we write $5 < 6$. The real number x is less than the real number $x - 1$. Hence we write $x < x - 1$.

9. If x is less than y, we denote this by writing $x < y$. If we write $x < y$ then x is less than y.

10. The statement $2 > 1$ is called an inequality. We shall also call the statement $1 < 2$ an inequality. For all real numbers x, y, we shall call the statement $x < y$ an inequality. This is because, as we shall later find out, both statements are equivalent.

11. **Not less than.** If x is not less than y, we denote this by writing $x \not< y$. If we write $x \not< y$ then x is not less than y.

1.6 Basic Properties of Less Than

1. In this section, we prove the Converse of Inequality theorem: $x < y$ if and only if $y > x$ for all real numbers x and y. Using this theorem, we rephrase the theorems we previously proved for greater than, such that they now use less than instead of greater than.

name	property
converse of inequality	$x < y$ if and only if $y > x$.
transitivity of inequality	If $x < y$ and $y < z$ then $x < z$.
addition of a constant	If $x < y$ then $x + w < y + w$.
addition of two inequalities	If $x < y$ and $w < z$ then $x + w < y + z$.
subtraction of a constant	If $x < y$ then $x - w < y - w$.
negation of an inequality	If $x < y$ then $-y < -x$.
subtraction of two inequalities	If $x < y$ and $w < z$ then $x - w < y - z$.
multiplication of an inequality by a positive constant	If $w > 0$ and $x < y$ and $wx < wy$.
multiplication of two inequalities	If $x, w > 0$, $x < y$ and $w < z$ then $xw < yz$.

2. Let w, x, y, z be real numbers. Let n be a positive integer.

3. If 5 is less than 6 then 6 is greater than 5. If 6 is greater than 5 then 5 is less than 6. This property is the converse of inequality. We prove it below.

4. $x < y$ if and only if $y > x$

    (Converse of Inequality)

    Proof We are given that $x < y$. Thus $x - y$ is negative. Hence $-(x - y)$ is positive. Observe that $-(x - y) = y - x$. Thus $y - x$ is positive. Therefore $y > x$. Conversely, let $y > x$. Thus $y - x$ is positive. Hence $-(y - x)$ is negative. Observe that $-(y - x) = x - y$. Thus $x - y$ is negative. Therefore $x < y$. ∎

5. x is negative if and only if $x < 0$.

Proof We know that x is negative. Thus $-x$ is positive. Hence $0-x$ is positive. It follows that $0 > x$. Therefore $x < 0$. Conversely, let $x < 0$. Then $0 > x$. Hence $0 - x$ is positive. Thus $-x$ is positive. Therefore x is negative. ∎

6   If $x < y$ and $y < z$ then $x < z$.

   (Transitivity of Inequality)

   Proof We are given that $x < y$ and $y < z$. Thus $y > x$ and $z > y$. From the transitivity of greater than we obtain $z > x$. Therefore $x < z$. ∎

7   If $x < y$ then $x + w < y + w$.

   (Addition of a Constant)

   Proof We are given that $x < y$. Thus $y > x$. Add w to both sides. Thus $y + w > x + w$. Therefore $x + w < y + w$. ∎

8   If $x < y$ and $w < z$ then $x + w < y + z$.

   (Addition of Two Inequalities)

   Proof We are given that $x < y$ and $w < z$. Thus $y > x$ and $z > w$. Add both inequalities together. Hence $y + z > x + w$. Therefore $x + w < y + z$. ∎

9   If $x < y$ then $x - w < y - w$.

   (Subtraction of a Constant)

   Proof

   We are given that $x < y$. Thus $y > x$. Hence $y - w > x - w$. It follows that $x - w < y - w$. ∎

10   If $x < y$ then $-y < -x$.

   (Negation of an Inequality)

   Proof We are given that $x < y$. Thus $y > x$. Hence $-x > -y$. It follows that $-y < -x$. ∎

11   If $x < y$ and $w < z$ then $x - z < y - w$.

   (Subtraction of Two Inequalities)

Proof We are given that $x < y$ and $w < z$. Thus $y > x$ and $z > w$. Hence $y - w > x - z$. It follows that $x - z < y - w$. ∎

12 If $w > 0$ and $x < y$ then $xw < yw$

(Multiplication of Inequality by a Positive Constant)

Proof We are given that $x < y$. Thus $y > x$. Multiply both sides by w. Hence $yw > xw$. Therefore $xw < yw$. ∎

13 If $x, w > 0$, $x < y$ and $w < z$ then $xw < yz$.

(Multiplication of Two Inequalities)

Proof We are given that $x < y$ and $w < z$. Thus $y > x$ and $z > w$. Multiply both inequalities together. Hence $yz > xw$. Therefore $xw < yz$. ∎

14 If $x > 0$ and $x < y$ then $x^n < y^n$.

(Raising Both Sides of an Inequality to a Power)

Proof We are given that $x < y$. Thus $y > x$. Hence $y^n > x^n$. Therefore $x^n < y^n$. ∎

15 $x < y$ or $y < x$ or $x = y$.

Proof Due to trichotomy, $y > x$ or $x > y$ or $y = x$. Thus $x < y$ or $y < x$ or $x = y$. ∎

Exercises 1.6

Let w, x, y, z be real numbers. Let n, m be positive integers. Prove the following.

1 If
$$x_1 < y_1,\ x_2 < y_2, \ldots,\ x_n < y_n$$
then
$$x_1 + x_2 + \cdots + x_n < y_1 + y_2 + \cdots + y_n.$$

(Addition of Arbitrary Number of Inequalities)

2 If $x_1 < y_1,\ x_2 < y_2, \ldots,\ x_n < y_n$ then $x_1 x_2 \cdots x_n < y_1 y_2 \cdots y_n$.

(Multiplication of Arbitrary Number of Inequalities)

3 If $x < 0$ then $\dfrac{1}{x} < 0$

4 If $w > 0$ and $x < y$ then $\dfrac{x}{w} < \dfrac{y}{w}$

   (Division of Inequality by a Constant)

5 If $x > 0$ and $x < y$ then $\dfrac{1}{y} < \dfrac{1}{x}$

   (Reciprocal of Inequality)

6 If $w, x > 0$, $x < y$ and $w < z$ then $\dfrac{x}{z} < \dfrac{y}{w}$.

   (Division of Two Inequalities)

1.7 Greater Than and Less Than

1 In this section, we prove some theorems that use both greater than and less than:

   1. $x < y$ if and only if $-x > -y$
   2. If $w < 0$ and $x < y$ then $xw > yw$
   3. $x \not< y$ if and only if $y \not> x$
   4. $x > y$ or $x < y$ or $x = y$.

2 Let w, x, y be real numbers.

3 $x < y$ if and only if $-x > -y$

   (Multiplying an Inequality by Minus One)

   Proof We are given that $x < y$. Thus $y > x$. Negating the inequality $y > x$ we obtain $-x > -y$.

   Conversely, let $-x > -y$. Negating the inequality $-x > -y$ we obtain $-(-y) > -(-x)$. Hence $y > x$. Thus $x < y$. ∎

4 The summary of the above result is that we can multiply an inequality by -1 provided we reverse the sign of the inequality afterwards. Thus 'greater than' changes to 'less than' and vice versa.

5 If $w < 0$ and $x < y$ then $xw > yw$

   (Multiplication of Inequality by a Negative Constant)

Proof We are given that $x < y$. Since w is negative, $-w$ is positive. Multiply both sides of $x < y$ by $-w$. Hence $-xw < -yw$. Multiply $-xw < -yw$ by -1. Therefore $xw > yw$. ∎

6 $x \not< y$ if and only if $y \not> x$

(Converse of Inequality)

Proof This is simply the contrapositive of $x < y$ if and only if $y > x$. ∎

7 $x > 0$ or $x < 0$ or $x = 0$

Proof Let x be a real number. Due to trichotomy, $x > 0$ or $0 > x$ or $x = 0$. Thus $x > 0$ or $x < 0$ or $x = 0$. ∎

8 Each real number is positive, negative or zero.

Proof Let x be a real number. Thus $x > 0$ or $x < 0$ or $x = 0$. Hence, x is positive or x is negative or x is zero. ∎

9 $x > y$ or $x < y$ or $x = y$.

Proof Consider the real number $x - y$. Due to the law of trichotomy, $x - y > 0$ or $x - y < 0$ or $x - y = 0$. Add y to both sides of each inequality/equality. Hence $x > y$ or $x < y$ or $x = y$. ∎

CHAPTER 2
Non-Strict Inequality

2.1 Nonnegative Real Number

1. In this section, we define nonnegative real number. Then we define the greater than or equal to relation and its notation.

2. **Definition of nonnegative real number.** A real number is a nonnegative real number if and only if it is not negative. Examples of such numbers are $0, 1, 2, 3, 3 + 1/2$.

   (Definition of Nonnegative Real Number)

3. If x is a nonnegative real number we say x is *nonnegative*. If we say x is nonnegative then x is a nonnegative real number.

4. A real number is a nonnegative real number if and only if it is positive or zero.

   Proof Let x be a real number. If x is nonnegative then x is not negative. Due to trichotomy x is negative or positive or zero. Since the first case is eliminated it follows that x is positive or zero.

   Conversely, suppose x is positive or zero. Thus it is not the case that x is negative. Hence x is not negative. Consequently, x is nonnegative. ∎

5. **Greater than or equal to.** We say x is *greater than or equal to* y if and only if $x - y$ is nonnegative.

6. **Notation of greater than or equal to.** If x is greater than or equal to y, we denote this by writing $x \geq y$. If we write $x \geq y$ then x is greater than or equal to y.

7. The statements $2 > 1$ and $3 < 5$ are called inequalities. We shall also call the statement $2 \geq 1$ an inequality. For all real numbers x, y, we shall call the statement $x \geq y$ an inequality.

2.2 Basic Properties of Greater Than or Equal To

1. In this section, we prove some properties of the greater than or equal to relation. The table below shows some properties we prove.

name	property
reflexivity	$x \geq x$
antisymmetry	If $x \geq y$ and $y \geq x$ then $x = y$.
totality	$x \geq y$ or $y \geq x$.
transitivity of dissimilar inequalities	If $x > y$ and $y \geq z$ then $x > z$.
transitivity of inequality	If $x \geq y$ and $y \geq z$ then $x \geq z$.
addition of two inequalities	If $x \geq y$ and $w \geq z$ then $x + w \geq y + z$.
multiplication of inequality by a nonnegative constant	If $w \geq 0$ and $x \geq y$ then $xw \geq yw$

2. Let w, x, y, z be real numbers. Let n be a positive integer.

3. $x \geq y$ if and only if $x > y$ or $x = y$

   Proof If $x \geq y$ then x is greater than y or x is equal to y. Thus $x > y$ or $x = y$. Conversely, if $x > y$ or $x = y$ then x is greater than y or x is equal to y. Therefore $x \geq y$. ∎

4. $x \geq y$ if and only if $x \not< y$

   (Greater Than or Equal To is Equivalent to Not Less Than)

   Proof Consider the real numbers x, y. The law of trichotomy implies that $x > y$ or $x = y$ or $x < y$.

   If $x \geq y$ then $x > y$ or $x = y$. Hence it is not the case that $x < y$. Thus x is not less than y. Therefore $x \not< y$.

   Conversely, if $x \not< y$ then x is not less than y. The only two remaining cases given by trichotomy are $x > y$ and $x = y$. Hence $x > y$ or $x = y$. Therefore $x \geq y$. ∎

5. $5 \geq 4$

2.2 Basic Properties of Greater Than or Equal To

Proof Since $5 > 4$ it follows that $5 > 4$ or $5 = 4$. Hence $5 \geq 4$. ∎

6   If $x > y$ then $x \geq y$.

Proof Since $x > y$, it follows that $x > y$ or $x = y$. Hence $x \geq y$. ∎

7   $5 \geq 5$

Proof Since $5 = 5$ it follows that $5 = 5$ or $5 > 5$. Hence $5 \geq 5$. ∎

8   $x \geq x$

(Reflexivity of Inequality)

Proof Since $x = x$ it follows that $x = x$ or $x > x$. Hence $x \geq x$. ∎

9   If $x \geq y$ and $y \geq x$ then $x = y$.

Proof Since $x \geq y$ it follows that $x \not< y$. Also, since $y \geq x$ it follows that $x \not< y$. The law of trichotomy gives that $x < y$ or $x > y$ or $x = y$. The first two cases have been eliminated. Hence the only case remaining is $x = y$. ∎

10   $x \geq y$ or $y \geq x$

(Totality of Greater Than or Equal To)

Proof Due to trichotomy, $x > y$ or $y = x$ or $y > x$. The last two cases imply $y \geq x$. Hence $x > y$ or $y \geq x$. The case $x > y$ implies $x \geq y$. Thus $x \geq y$ or $y \geq x$. ∎

11   If $x > y$ and $y \geq z$ then $x > z$.

(Transitivity of Dissimilar Inequalities)

Proof

We are given that $y \geq z$. Thus $y > z$ or $y = z$. Consider the case $y > z$. Since $x > y$ and $y > z$, it follows that $x > z$. Consider the case $y = z$. By subsisting z for y in the inequality $x > y$, we get $x > z$. Therefore in both cases $x > z$. ∎

12   If $x \geq y$ and $y \geq z$ then $x \geq z$.

(Transitivity of Inequality)

Proof

We are given that $x \geq y$. Thus $x > y$ or $x = y$. Consider the case $x > y$. Since $x > y$ and $y \geq z$, it follows that $x > z$. Hence $x \geq z$. Consider the case $x = y$. By substituting x for y in the inequality $y \geq z$, we get $x \geq z$. Therefore in both cases $x \geq z$. ∎

13 If $x \geq y$ then $x + w \geq y + w$.

(Addition of Constant to Inequalities)

Proof

We are given that $x \geq y$. Thus $x > y$ or $x = y$. Consider the case $x > y$. Adding w to both sides of the inequality we get $x + w > w + z$. Hence $x + w \geq w + z$. Consider the case $x = y$. Adding w to both sides of the equation, we get $x + w = w + z$. Thus $x + w \geq y + w$. Therefore in both cases $x + w \geq y + w$. ∎

14 If $x > y$ and $w \geq z$ then $x + w > y + z$.

(Addition of Dissimilar Inequalities)

Proof

We are given that $w \geq z$. Thus $w > z$ or $w = z$. Consider the case $w > z$. Adding the inequalities $x > y$ and $w > z$, we obtain $x + w > y + z$. Consider the case $w = z$. Adding w to both sides of the inequality $x > y$, we get $x + w > y + w$. Substituting z for w in the RHS of the inequality, we obtain $x + w > y + z$. Therefore in both cases $x + w > y + z$. ∎

15 If $x \geq y$ and $w \geq z$ then $x + w \geq y + z$.

(Addition of Two Inequalities)

Proof

We are given that $x \geq y$. Thus $x > y$ or $x = y$. Consider the case $x > y$. Adding the inequalities $x > y$ and $w \geq z$, we obtain $x + w > y + z$. Hence $x + w \geq y + z$. Consider the case $x = y$. Adding x to both sides of the inequality $w \geq z$, we get $x + w \geq x + z$. Substituting y for x in the RHS of the inequality $x + w \geq x + z$, we obtain $x + w \geq y + z$. Therefore in both cases $x + w \geq y + z$. ∎

16 If $w \geq 0$ and $x \geq y$ then $xw \geq yw$

(Multiplication of Inequality by a Nonnegative Constant)

2.2 Basic Properties of Greater Than or Equal To

Proof We are given that $w \geq 0$. Thus $w > 0$ or $w = 0$. When $w = 0$ we know that $xw = 0$ and $yw = 0$. Thus $xw = yw$. Hence $xw \geq yw$.

When $w > 0$. We are given $x \geq y$. Thus $x > y$ or $x = y$. Consider the case $x > y$. Multiplying both sides of the inequality by w we get $xw \geq wy$. Hence $xw \geq yw$. Consider the case $x = y$. Multiplying both sides of the equation by w, we get $xw = yw$. Thus $xw \geq yw$. Therefore in both cases $xw \geq yw$. ∎

17   Let $y, z > 0$. If $x > y$ and $w \geq z$ then $xw > yz$.

(Multiplication of Dissimilar Inequalities)

Proof

We are given that $w \geq z$. Thus $w > z$ or $w = z$. Consider the case $w > z$. Multiplying the inequalities $x > y$ and $w > z$, we obtain $xw > yz$.

Consider the case $w = z$. Since $z > 0$, it follows that $w > 0$. Multiplying both sides of the inequality $x > y$ by w, we get $xw > yw$. Substituting z for w in the RHS of the inequality, we obtain $xw > yz$.

Therefore in both cases $xw > yz$. ∎

18   Let $y, z \geq 0$. If $x \geq y$ and $w \geq z$ then $xw \geq yz$.

(Multiplication of Two Inequalities)

Proof Multiply $x \geq y$ by w. We get $xw \geq yw$. Multiply $w \geq z$ by y. We get $yw \geq yz$. Apply transitivity to $xw \geq yw$ and $yw \geq yz$. We get $xw \geq yz$. ∎

19   **Not greater than or equal to.** If x is not greater than or equal to y, we denote this by writing $x \not\geq y$. If we write $x \not\geq y$ then x is not greater than or equal to y.

Exercises 2.2

Let w, x, y, z be real numbers. Let n, m be positive integers.

1   If $x \geq y$ then $x - w \geq y - w$.

(Subtraction of a Constant)

2   If $x \geq y$ and $w \geq z$ then $x - z \geq y - w$.

(Subtraction of Two Inequalities)

3 If
$$x_1 \geq y_1,\ x_2 \geq y_2, \ldots, x_n \geq y_n$$
then
$$x_1 + x_2 + \cdots + x_n \geq y_1 + y_2 + \cdots + y_n$$
(Addition of Arbitrary Number of Inequalities)

4 If $x \geq y$ and $y > z$ then $x > z$.
(Transitivity of Dissimilar Inequalities II)

5 If $x \geq y$ and $w > z$ then $x - z > y - w$.
(Subtraction of Dissimilar Inequalities I)

6 If $x > y$ and $w \geq z$ then $x - z > y - w$.
(Subtraction of Dissimilar Inequalities II)

7 If $w > 0$ and $x \geq y$ then $\dfrac{x}{w} \geq \dfrac{y}{w}$
(Division of Inequality by a Constant)

Exercises 2.2

Let w, x, y, z be real numbers and let n, m be positive integers.

1 If
$$x_1 \geq y_1,\ x_2 \geq y_2, \ldots, x_n \geq y_n$$
then
$$x_1 x_2 \cdots x_n \geq y_1 y_2 \cdots y_n$$
(Multiplication of Arbitrary Number of Inequalities)

2 Let $y_1 > 0$ and $y_2, y_3 \geq 0$. If
$$x_1 > y_1, \quad x_2 \geq y_2, \quad x_3 \geq y_3$$
then
$$x_1 x_2 x_3 > y_1 y_2 y_3$$
(Multiplication of Three Inequalities, with One Dissimilar)

3 If $y \geq 0$ and $x \geq y$ then $x^n \geq y^n$.

4 Let $y, z > 0$. If $x \geq y$ and $w > z$ then $\dfrac{x}{z} > \dfrac{y}{w}$.

   (Division of Dissimilar Inequalities I)

5 Let $y, z > 0$. If $x > y$ and $w \geq z$ then $\dfrac{x}{z} > \dfrac{y}{w}$.

   (Division of Dissimilar Inequalities II)

6 Let $y, z > 0$. If $x \geq y$ and $w \geq z$ then $\dfrac{x}{z} \geq \dfrac{y}{w}$.

   (Division of Inequalities)

2.3 Zero or Negative Real Number

1 In this section, we define the less than or equal to relation and its notation.

2 **Less than or equal to.** We say x is *less than or equal to* y if and only if $x - y$ is negative or zero.

3 **Notation of less than or equal to.** If x is less than or equal to y, we denote this by writing $x \leq y$. If we write $x \leq y$ then x is less than or equal to y.

4 The statements $2 > 1$, $3 < 5$ and $2 \geq 1$ are called inequalities. We shall also call the statement $2 \leq 1$ an inequality. For all real numbers x, y, we shall call the statement $x \leq y$ an inequality.

2.4 Basic Properties of Less Than or Equal To

1 In this section, we prove some properties of the less than or equal to relation. The table below shows some properties we prove.

name	property
reflexivity	$x \leq x$
antisymmetry	If $x \leq y$ and $y \leq x$ then $x = y$.
totality	$x \leq y$ or $y \leq x$.
transitivity of dissimilar inequalities	If $x < y$ and $y \leq z$ then $x < z$.
transitivity of inequality	If $x \leq y$ and $y \leq z$ then $x \leq z$.
addition of two inequalities	If $x \leq y$ and $w \leq z$ then $x + w \leq y + z$.
multiplication of inequality by a nonnegative constant	If $w \geq 0$ and $x \leq y$ then $xw \leq yw$

2   Let w, x, y, z be real numbers. Let n be a positive integer.

3   $x \leq y$ if and only if $y \geq x$

(Converse of Inequality)

Proof

We are given that $x \leq y$. Thus $x < y$ or $x = y$. Consider the case $x < y$. Due to converse of inequality, $y > x$. Thus $y \geq x$.

Consider the case $x = y$. Hence, $y = x$. Thus $y \geq x$. ■

4   $x \leq y$ if and only if $x < y$ or $x = y$

Proof   If $x \leq y$ then x is less than y or x is equal to y. Thus $x < y$ or $x = y$. Conversely, if $x < y$ or $x = y$ then x is less than y or x is equal to y. Therefore $x \leq y$. ■

5   $x \leq y$ if and only if $x \not> y$

(Less Than or Equal To is Equivalent to Not Greater Than)

Proof   From a previous result, $y \geq x$ if and only if $y \not< x$. Thus $x \leq y$ if and only if $y \not< x$. Hence $x \leq y$ if and only if $x \not> y$. ■

6   If $x < y$ then $x \leq y$.

2.4 Basic Properties of Less Than or Equal To

Proof Since $x < y$, it follows that $x < y$ or $x = y$. Hence $x \leq y$. ∎

7 $x \leq x$

(Reflexivity of Inequality)

Proof Since $x = x$ it follows that $x = x$ or $x < x$. Hence $x \leq x$. ∎

8 If $x \leq y$ and $y \leq x$ then $x = y$.

Proof From a previous result, if $y \geq x$ and $x \geq y$ then $y = x$. Thus, if $x \leq y$ and $y \leq x$ then $x = y$. ∎

9 $x \leq y$ or $y \leq x$

(Totality of Less Than or Equal To)

Proof From a previous result, $y \geq x$ or $x \geq y$. Thus $x \leq y$ or $y \leq x$. ∎

10 If $x < y$ and $y \leq z$ then $x < z$.

(Transitivity of Dissimilar Inequalities)

Proof proof ∎

11 If $x \leq y$ and $y \leq z$ then $x \leq z$.

(Transitivity of Inequality)

Proof From a previous result, if $z \geq y$ and $y \geq x$ then $z \geq x$. Thus, if $y \leq z$ and $x \leq y$ then $x \leq z$. Hence, if $x \leq y$ and $y \leq z$ then $x \leq z$. ∎

12 If $x \leq y$ then $x + w \leq y + w$.

(Addition of Constant to Inequalities)

Proof From a previous result, if $y \geq x$ then $y + w \geq x + w$. Thus, if $x \leq y$ then $x + w \leq y + w$. ∎

13 If $x < y$ and $w \leq z$ then $x + w < y + z$.

(Addition of Dissimilar Inequalities)

Proof

From a previous result, if $y > x$ and $z \geq w$ then $y + z > x + w$. Thus, if $x < y$ and $w \leq z$ then $x + w < y + z$. ∎

14   If $x \leq y$ and $w \leq z$ then $x + w \leq y + z$.

(Addition of Two Inequalities)

Proof   From a previous result, if $y \geq x$ and $z \geq w$ then $y + z \leq x + w$. Thus, if $x \leq y$ and $w \leq z$ then $x + w \leq y + z$. ∎

15   If $w \geq 0$ and $x \leq y$ then $xw \leq yw$

(Multiplication of Inequality by a Nonnegative Constant)

Proof   From a previous result, if $w \geq 0$ and $y \geq x$ then $yw \geq xw$. Thus, if $w \geq 0$ and $x \leq y$ then $xw \leq yw$. ∎

16   Let $x, w > 0$. If $x < y$ and $w \leq z$ then $xw < yz$.

(Multiplication of Dissimilar Inequalities)

Proof   From a previous result, if $y > x$ and $z \geq w$ then $yz > xw$. Thus, if $x < y$ and $w \leq z$ then $xw < yz$. ∎

17   Let $x, w \geq 0$. If $x \leq y$ and $w \leq z$ then $xw \leq yz$.

(Multiplication of Two Inequalities)

Proof   From a previous result, if $y \geq x$ and $z \geq w$ then $yz \geq xw$. Thus, if $x \leq y$ and $w \leq z$ then $xw \leq yz$. ∎

18   **Not less than or equal to.** If x is not less than or equal to y, we denote this by writing $x \not\leq y$. If we write $x \not\leq y$ then x is not less than or equal to y.

Exercises 2.4

1 Let w, x, y, z be real numbers. Let n, m be positive integers.

  a If $x \leq y$ then $x - w \leq y - w$.
(Subtraction of a Constant)

  b If $x \leq y$ and $w \leq z$ then $x - z \leq y - w$.
(Subtraction of Two Inequalities)

  c If $x \leq y$ and $w < z$ then $x - z < y - w$.
(Subtraction of Dissimilar Inequalities)

  d If

$$x_1 \leq y_1, \; x_2 \leq y_2, \ldots, x_n \leq y_n$$

then

$$x_1 + x_2 + \cdots + x_n \leq y_1 + y_2 + \cdots + y_n$$

(Addition of Arbitrary Number of Inequalities)

  e If $x \leq y$ and $y < z$ then $x < z$.
(Transitivity of Dissimilar Inequalities II)

  f If $x \leq y$ and $w < z$ then $x - z < y - w$.
(Subtraction of Dissimilar Inequalities I)

  g If $x < y$ and $w \leq z$ then $x - z < y - w$.
(Subtraction of Dissimilar Inequalities II)

  h If $w > 0$ and $x \leq y$ then $\dfrac{x}{w} \leq \dfrac{y}{w}$
(Division of Inequality by a Constant)

2 Let w, x, y, z be real numbers and let n, m be positive integers.

  a If

$$x_1 \leq y_1, \; x_2 \leq y_2, \ldots, x_n \leq y_n$$

then

$$x_1 x_2 \cdots x_n \leq y_1 y_2 \cdots y_n$$

(Multiplication of Arbitrary Number of Inequalities)

b Let $x_1 > 0$ and $x_2, x_3 \geq 0$. If

$$x_1 < y_1, \quad x_2 \leq y_2, \quad x_3 \leq y_3$$

then

$$x_1 x_2 x_3 < y_1 y_2 y_3$$

(Multiplication of Three Inequalities, with One Dissimilar)

c If $x \geq 0$ and $x \leq y$ then $x^n \leq y^n$.

d If $x \geq 0$ and $x \leq y$ then $x^n \leq y^n$.

e Let $x, w > 0$. If $x \leq y$ and $w < z$ then $\dfrac{x}{z} < \dfrac{y}{w}$.
(Division of Dissimilar Inequalities I)

f Let $x, w > 0$. If $x < y$ and $w \leq z$ then $\dfrac{x}{z} < \dfrac{y}{w}$.
(Division of Dissimilar Inequalities II)

g Let $x, w > 0$. If $x \leq y$ and $w \leq z$ then $\dfrac{x}{z} \leq \dfrac{y}{w}$.
(Division of Inequalities)

2.5 Definition of Terms

1 In this section, we define the following terms:

- inequality relation
- inequality
- strict and non-strict inequality
- similar and dissimilar inequalities

Then we compare the inequality and the equality relations.

2 **Inequality relation.** An inequality relation is one of the following:

- greater than
- less than
- greater than or equal to
- less than or equal to

3. **Inequality.** An inequality is a statement that compares two real numbers using an inequality relation. Examples of inequalities are

- a greater than inequality such as $5 > 4$
- a less than inequality such as $4 < 5$
- a greater than or equal to inequality such as $5 \geq 4$
- a less than or equal to inequality such as $4 \leq 5$

4. **Strict and non-strict inequality.** A strict inequality relation is one of the following

- greater than
- less than

A strict inequality is an inequality using one of the above relations. Examples are $5 > 4$ and $4 < 5$.

5. A non-strict inequality relation is one of the following

- greater than or equal to
- less than or equal to

A non-strict inequality is an inequality using one of the above relations. Examples are $5 \geq 4$ and $4 \leq 5$.

6. **Similar and dissimilar inequalities.** We say the inequalities $5 > 4$ and $6 > 3$ are similar. And we say the inequalities $3 \geq 2$ and $7 \geq 3$ are similar. Two inequalities are *similar* if they are both strict inequalities or both non-strict inequalities.

We say the inequalities $5 > 4$ and $6 \geq 3$ are dissimilar. Two inequalities are *dissimilar* if one is a strict inequality and the other is a non-strict inequality.

By the above definition the inequalities $5 > 4$ and $3 < 6$ are similar. While $5 > 4$ and $3 \leq 6$ are dissimilar inequalities.

7. **Comparing inequality and equality relation.** Another kind of relation we have encountered is the equality relation. Using the equality relation we can compare $4 \cdot 6$ and $5^2 - 1$:
$$4 \cdot 6 = 5^2 - 1$$

The equality relation and inequality relations have much in common. Particularly, the non-strict inequality relations have similar properties with the equality relation. We explore this similarity below by comparing the equality relation and the greater than or equal to relation.

8 Let a, b, c, d be integers.

property	equality	greater than or equal to
reflexivity	$a = a$	$a \geq a$
transitivity	If $a = b$ and $b = c$ then $a = c$	If $a \geq b$ and $b \geq c$ then $a \geq c$
addition of a constant	If $a = b$ then $a + c = b + c$	If $a \geq b$ then $a + c \geq b + c$
multiplying by a constant	If $a = b$ then $ac = bc$	If $c \geq 0$ and $a \geq b$ then $ac \geq bc$

2.6 Integer inequality

1 Inequalities that consist of integers have some special properties. We discuss some of such properties in this section.

2 **Integers and Greater Than.** Let n, m be integers.

3 If $n > 0$ then $n \geq 1$.

  Proof We are given that $n > 0$. Hence n is a positive integer. Thus $n = 1$ or n is greater than 1. Therefore $n = 1$ or $n > 1$. Hence $n \geq 1$. ∎

4 If $n > m$ then $n \geq m + 1$.

  Proof We are given that $n > m$. Subtract m from both sides. Thus $n - m > 0$. Since $n - m$ is an integer, it follows that $n - m \geq 1$. Therefore, $n \geq m + 1$. ∎

5 **Examples of integer inequalities.** Here, we prove some integer inequalities:

  - If $n > 1$ then $n^2 \geq 2n$
  - If $n > 1$ then $2n > n + 1$
  - If $n > 1$ then $n^2 > n + 1$
  - If $n \geq 3$ then $n^3 > (n + 1)^2$
  - If $n < -2$ then $n(n - 1) > 6$

2.6 Integer inequality

6 Let n be a positive integer.

7 If $n > 1$ then $n^2 \geq 2n$

Proof We are given that $n > 1$. Thus $n \geq 2$. Multiply both sides by n. We get $n^2 \geq 2n$. ∎

8 If $n > 1$ then $2n > n + 1$

Proof

We are given that $n > 1$. Add n to both sides. We obtain $n + n > n + 1$. Hence $2n > n + 1$. ∎

9 If $n > 1$ then $n^2 > n + 1$.

Proof We know from the previous two results that $n^2 \geq 2n$ and $2n > n + 1$. Applying transitivity to these two inequalities we obtain $n^2 > n + 1$. ∎

10 If $n \geq 3$ then $n^3 > (n + 1)^2$.

Proof

$$n \geq 3$$
$$n - 1 \geq 2$$

multiply the two previous inequalities together

$$n(n - 1) \geq 3 \cdot 2$$
$$n(n - 1) \geq 6$$
$$n(n - 1) > 5$$
$$n^2 - n > 5$$

Subtract 2 from both sides

$$n^2 - n - 2 > 3$$

multiply both sides by n

$$n(n^2 - n - 2) > 3n$$

Since $3n > 1$ we may write

$$n(n^2 - n - 2) > 1$$
$$n^3 - n^2 - 2n > 1$$
$$n^3 > 1 + n^2 + 2n$$
$$n^3 > (n + 1)^2$$

∎

11 If $n < -2$ then $n(n - 1) > 6$

Proof We are given that

$$n < -2$$

multiply both sides by -1. Since the factor is negative then we would have to reverse the sign of the inequality

$$-n > 2 \qquad (1)$$

add 1 to both sides

$$1 - n > 1 + 2$$
$$1 - n > 3 \qquad (2)$$

multiply the above inequality and (1)

$$-n(1 - n) > 2 \cdot 3$$
$$n(n - 1) > 6$$

∎

12 **Factorial versus square.** Here, we prove that if an integer is greater than 3 then its factorial is greater than its square. That is, for all integers $n > 3$,

$$n! > n^2.$$

Before proving this theorem, we prove some lemmas.

13 If $n > 1$ then $n(n + 1) > (n + 2)$

Proof

$$n > 1$$
$$n \geq 2$$

square both sides

$$n^2 \geq 4$$
$$n^2 > 2$$

add n to both sides

$$n^2 + n > n + 2$$
$$n(n + 1) > n + 2$$

∎

14 If $n > 3$ then
$$n(n - 1)(n - 2)(n - 3) > n^2$$

Proof First, we prove that $(n - 2)(n - 1) > n$.

$$n > 3$$
$$n - 2 > 1$$

we can apply a previous result

$$(n - 2)(n - 2 + 1) > (n - 2 + 2)$$
$$(n - 2)(n - 1) > n$$

Next, we prove that $n(n - 3) \geq n$.

$$n > 3$$
$$n \geq 4$$
$$n - 3 \geq 1$$

multiply both sides by n

$$n(n - 3) \geq n$$

Now, we have that

$$(n - 2)(n - 1) > n$$

and
$$n(n-3) \geq n$$
multiply both inequalities together
$$(n-2)(n-1) \cdot n(n-3) > n \cdot n$$
$$n(n-1)(n-2)(n-3) > n^2$$

∎

15 If $n > 3$ then
$$n! > n^2$$

Proof

$$n > 3$$
$$n \geq 4$$
$$n - 4 \geq 0$$
$$(n-4)! \geq 1$$

multiply both sides by $n(n-1)(n-2)(n-3)$

$$n(n-1)(n-2)(n-3)(n-4)! \geq n(n-1)(n-2)(n-3)$$
$$n! \geq n(n-1)(n-2)(n-3)$$

from a previous result, we know that
$$n(n-1)(n-2)(n-3) > n^2$$

due to transitivity of inequality, we have that
$$n! > n^2$$

∎

Exercises 2.6

1 Prove the following for all positive integers n, m.

   a If $n < m$ then $n = m - 1$ or $n < m - 1$.

   b $n > 0$ if and only if $n \geq 1$.

c $n > m$ if and only if $n \geq m + 1$.

d $n < 0$ if and only if $n \leq -1$.

e $n < m$ if and only if $n \leq m - 1$.

2 Let n be an integer greater than 5. Prove the following

a $n(n-5) > n$

b $(n-1)(n-4) > n$

c $(n-2)(n-3) > n$

d $n(n-1)\cdots(n-5) > n^3$

e $n! > n(n-1)\cdots(n-5)$

f $n! > n^3$

2.7 Bernoulli Inequality

1 In this section, we prove the Bernoulli inequality: for all real numbers a and for all integers n, if $1 + a > 0$ and $n \geq 0$ then

$$(1+a)^n > 1 + an.$$

Before proving this inequality, we prove some of its corollaries.

2 Let a, n be integers.

3 If $n \geq 0$ then $2^n \geq n + 1$

Proof

$$1 \geq 1$$
$$2 \geq 1$$
$$2^2 \geq 1$$
$$2^3 \geq 1$$
$$\vdots$$
$$2^{n-2} \geq 1$$
$$2^{n-1} \geq 1$$

If we add up all the inequalities we get

$$1 + 2 + 2^2 + 2^3 + \cdots + 2^{n-2} + 2^{n-1} \geq 1 + 1 + 1 + 1 + \cdots + 1 + 1$$

there are n terms on either side. Thus

$$2^n - 1 \geq n$$

add 1 to both sides

$$2^n \geq n + 1$$

∎

4 If $n \geq 0$ then $3^n \geq 2n + 1$

Proof

$$1 \geq 1$$
$$3 \geq 1$$
$$3^2 \geq 1$$
$$3^3 \geq 1$$
$$3^4 \geq 1$$
$$\vdots$$
$$3^{n-2} \geq 1$$
$$3^{n-1} \geq 1$$

If we add up all the inequalities we get

$$1 + 3 + 3^2 + 3^3 + \cdots + 3^{n-2} + 3^{n-1} \geq 1 + 1 + 1 + 1 + \cdots + 1 + 1$$

there are n terms on either side. Thus

$$\frac{3^n - 1}{2} \geq n$$
$$3^n - 1 \geq 2n$$

add 1 to both sides

$$3^n \geq 2n + 1$$

∎

5 If $n \geq 0$ and $a \geq 1$ then $a^n \geq (a-1)n + 1$

Proof

$$1 \geq 1$$
$$a \geq 1$$
$$a^2 \geq 1$$
$$a^3 \geq 1$$
$$\vdots$$
$$a^{n-2} \geq 1$$
$$a^{n-1} \geq 1$$

If we add up all the inequalities we get

$$1 + a + a^2 + a^3 + \cdots + a^{n-2} + a^{n-1} \geq 1 + 1 + 1 + 1 + \cdots + 1 + 1$$

there are n terms on the RHS. Thus

$$1 + a + a^2 + a^3 + \cdots + a^{n-2} + a^{n-1} \geq n$$

multiply both sides by $a - 1$

$$(a-1)(1 + a + a^2 + a^3 + \cdots + a^{n-2} + a^{n-1}) \geq (a-1)n$$
$$a^n - 1 \geq (a-1)n$$

add 1 to both sides

$$a^n \geq (a-1)n + 1$$

∎

6 If $1 + a > 0$ and $n \geq 0$ then $(1+a)^n > 1 + an$

(Bernoulli Inequality)

Proof We are given that $1+a > 0$. From a previous result $(1+a)^n \geq an+1$. ∎

CHAPTER 3
Chained Inequality

3.1 Introduction

1. In this section, we define the following concepts: chained inequality, chain and split.

2. **Definition of chained inequality.** A short way of writing $1 < 2$ and $2 < 3$ is
$$1 < 2 < 3.$$
Similarly, we may write $5 > 4$ and $4 \geq 2$ as
$$5 > 4 \geq 2.$$

In general, we may use this notation to write a pair of inequalities that share a number. For example, the inequalities $10 \leq 11$ and $11 \leq 12$ share the number 11. Thus we may write $10 \leq 11 \leq 12$. Here are some more examples: we write

- $x \leq y < z$ if and only if $x \leq y$ and $y < z$.
- $x > y > z$ if and only if $x > y$ and $y > z$.
- $x \geq y > z$ if and only if $x \geq y$ and $y > z$.
- $x \geq y \geq z$ if and only if $x \geq y$ and $y \geq z$.

An inequality such as one of the types above is called a *chained inequality*.

3. **Definition of chain and split.** We *chain* two inequalities $x > y$ and $y > z$ by writing $x > y > z$. We *split* a chained inequality $x > y > z$ by writing $x > y$ and $y > z$.

3.2 Strict Chained Inequality

1. In this section, we prove some properties of chained inequalities, which are summarized in the table below.

name	property
converse of a chained inequality	$a < x < b$ if and only if $b > x > a$.
addition of a constant	If $a < x < b$ then $a + c < x + c < b + c$.
multiplying by a positive constant	If $c > 0$ and $a < x < b$ then $ac < xc < bc$.
addition of two chained inequalities	If $a < x < b$ and $c < y < d$ then $a + c < x + y < b + d$.
subtraction of two chained inequalities	If $a < x < b$ and $c < y < d$ then $a - d < x - y < b - c$.
multiplication of two chained inequalities	Let $a > 0$ and $c > 0$. If $a < x < b$ and $c < y < d$ then $ac < xy < bd$.
division of two chained inequalities	Let $a, c > 0$. If $a < x < b$ and $c < y < d$ then $\frac{a}{d} < \frac{x}{y} < \frac{b}{c}$.

2. Let a, b, c, d, x, y be real numbers.

3. $a < x < b$ if and only if $b > x > a$.

   (Converse of a Chained Inequality)

   Proof We are given that

   $$a < x < b$$

   we can split the above chained inequality

   $$a < x \qquad (1)$$
   $$x < b \qquad (2)$$

   The converse of (1) is

   $$x > a \qquad (3)$$

The converse of (2) is

$$b > x \qquad (4)$$

Chain (4) and (3)

$$b > x > a$$

Conversely, we are given that

$$b > x > a$$

we can split the above chained inequality

$$b > x \qquad (5)$$
$$x > a \qquad (6)$$

The converse of (5) is

$$x < b \qquad (7)$$

The converse of (6) is

$$a < x \qquad (8)$$

Chain (8) and (7)

$$a < x < b$$

∎

4 If $a < x < b$ then $a + c < x + c < b + c$.
(Addition of a Constant)

Proof We are given that

$$a < x < b$$

we can split the above chained inequality

$$a < x \qquad (1)$$
$$x < b \qquad (2)$$

add c to both sides of (1)

$$a + c < x + c \tag{3}$$

add c to both sides of (2)

$$x + c < b + c \tag{4}$$

chain (3) and (4)

$$a + c < x + c < b + c$$

∎

5 If $c > 0$ and $a < x < b$ then $ac < xc < bc$.
(Multiplying by a Positive Constant)

Proof We are given that

$$a < x < b$$

we can split the above chained inequality

$$a < x \tag{1}$$
$$x < b \tag{2}$$

multiply both sides of (1) by c

$$ac < xc \tag{3}$$

multiply both sides of (2) by c

$$xc < bc \tag{4}$$

chain (3) and (4)

$$ac < xc < bc$$

∎

6 If $a < x < b$ and $c < y < d$ then $a + c < x + y < b + d$.
(Addition of Two Chained Inequalities)

Proof We are given that
$$a < x < b$$
we can split the above chained inequality
$$a < x \qquad (1)$$
$$x < b \qquad (2)$$
We are also given that
$$c < y < d$$
we can split the above chained inequality
$$c < y \qquad (3)$$
$$y < d \qquad (4)$$
Add (1) and (3)
$$a + c < x + y \qquad (5)$$
Add (2) and (4)
$$x + y < b + d \qquad (6)$$
chain (5) and (6)
$$a + c < x + y < b + d$$

∎

7. If $a < x < b$ and $c < y < d$ then $a - d < x - y < b - c$. (Subtraction of Two Chained Inequalities)

Proof We are given that
$$a < x < b \qquad (1)$$
$$c < y < d \qquad (2)$$
Multiply through (2) by -1
$$-c > -y > -d$$
$$-d < -y < -c \qquad (3)$$
Add (1) and (3)
$$a - d < x - y < b - c$$

∎

8 Let $a, c > 0$. If $a < x < b$ and $c < y < d$ then $ac < xy < bd$. (Multiplication of Two Chained Inequalities)

Proof We are given that

$$a < x < b$$

we can split the above chained inequality

$$a < x \qquad (1)$$
$$x < b \qquad (2)$$

We are also given that

$$c < y < d$$

we can split the above chained inequality

$$c < y \qquad (3)$$
$$y < d \qquad (4)$$

Multiply (1) and (3)

$$ac < xy \qquad (5)$$

Multiply (2) and (4)

$$xy < bd \qquad (6)$$

chain (5) and (6)

$$ac < xy < bd$$

∎

9 Let $a, c > 0$. If $a < x < b$ and $c < y < d$ then $\dfrac{a}{d} < \dfrac{x}{y} < \dfrac{b}{c}$. (Division of Two Chained Inequalities)

Proof We are given that

$$a < x < b$$
$$c < y < d$$

we can split each of the above chained inequality

$$a < x \quad (1)$$
$$x < b \quad (2)$$
$$c < y \quad (3)$$
$$y < d \quad (4)$$

Since $0 < c$ and $c < y$, by transitivity, $0 < y$. Hence $c, y > 0$. Divide (1) by (4)

$$\frac{a}{d} < \frac{x}{y}$$

Divide (2) by (3)

$$\frac{x}{y} < \frac{b}{c}$$

Chain the last two inequalities together

$$\frac{a}{d} < \frac{x}{y} < \frac{b}{c}$$

∎

Exercises 3.2

Let a, b, c, d, x, y be real numbers. Let $a_1, \ldots, a_n$, $b_1, \ldots, b_n$ and $x_1, \ldots, x_n$ be real numbers. Prove the following.

1 **a** If $a > x > b$ then $a + c > x + c > b + c$.
   (Addition of a Constant)

   b If $c > 0$ and $a > x > b$ then $ac > xc > bc$.
   (Multiplying by a Positive Constant)

   c If $a > x > b$ and $c > y > d$ then $a + c > x + y > b + d$.
   (Addition of Two Chained Inequalities)

   d If $a > x > b$ and $c > y > d$ then $a - d > x - y > b - c$.
   (Subtraction of Two Chained Inequalities)

   e Let $b, d > 0$. If $a > x > b$ and $c > y > d$ then $ac > xy > bd$.
   (Multiplication of Two Chained Inequalities)

f Let $b, d > 0$. If $a > x > b$ and $c > y > d$ then $\dfrac{a}{d} > \dfrac{x}{y} > \dfrac{b}{c}$.
(Division of Two Chained Inequalities)

2 a If $c > 0$ and $a < x < b$ then $a - c < x < b + c$.
(Superset of a Chained Inequality)

b If $c < 0$ and $a < x < b$ then $ac > xc > bc$.
(Multiplying by a Negative Constant)

c If
$$a_1 < x_1 < b_1, \ a_2 < x_2 < b_2, \ldots, a_n < x_n < b_n$$
then
$$a_1 + \cdots + a_n < x_1 + \cdots + x_n < b_1 + \cdots + b_n.$$
(Addition of Arbitrary Number of Chained Inequalities)

d Let $a_1, \ldots, a_n > 0$. If
$$a_1 < x_1 < b_1, \ a_2 < x_2 < b_2, \ldots, a_n < x_n < b_n$$
then
$$a_1 \cdots a_n < x_1 \cdots x_n < b_1 \cdots b_n.$$
(Multiplication of Arbitrary Number of Chained Inequalities)

e If $a > 0$ and $a < x < b$ then $a^n < x^n < b^n$.
(Power of a Chained Inequality)

3 a If $c > 0$ and $a > x > b$ then $a + c > x > b - c$.
(Superset of a Chained Inequality)

b If $c < 0$ and $a > x > b$ then $ac < xc < bc$.
(Multiplying by a Negative Constant)

c If
$$a_1 > x_1 > b_1, \ a_2 > x_2 > b_2, \ldots, a_n > x_n > b_n$$
then
$$a_1 + \cdots + a_n > x_1 + \cdots + x_n > b_1 + \cdots + b_n.$$
(Addition of Arbitrary Number of Chained Inequalities)

d Let $b_1, \ldots, b_n > 0$. If
$$a_1 > x_1 > b_1, \; a_2 > x_2 > b_2, \ldots, a_n > x_n > b_n$$
then
$$a_1 \cdots a_n > x_1 \cdots x_n > b_1 \cdots b_n.$$
(Multiplication of Arbitrary Number of Chained Inequalities)

e If $b > 0$ and $a > x > b$ then $a^n > x^n > b^n$.
(Power of a Chained Inequality)

3.3 Non-Strict Chained Inequality

1 In this section, we prove some properties of chained inequalities, which are summarized in the table below.

name	property
converse of a chained inequality	$a \leq x \leq b$ if and only if $b \geq x \geq a$.
addition of a constant	If $a \leq x \leq b$ then $a + c \leq x + c \leq b + c$.
multiplying by a nonnegative constant	If $c \geq 0$ and $a \leq x \leq b$ then $ac \leq xc \leq bc$.
addition of two chained inequalities	If $a \leq x \leq b$ and $c \leq y \leq d$ then $a + c \leq x + y \leq b + d$.
subtraction of two chained inequalities	If $a \leq x \leq b$ and $c \leq y \leq d$ then $a - d \leq x - y \leq b - c$.
multiplication of two chained inequalities	Let $a \geq 0$ and $c \geq 0$. If $a \leq x \leq b$ and $c \leq y \leq d$ then $ac \leq xy \leq bd$.
division of two chained inequalities	Let $a, c > 0$. If $a \leq x \leq b$ and $c \leq y \leq d$ then $\dfrac{a}{d} \leq \dfrac{x}{y} \leq \dfrac{b}{c}$.

2 Let a, b, c, d, x, y be real numbers.

3 $a \leq x \leq b$ if and only if $b \geq x \geq a$.

(Converse of a Chained Inequality)

Proof We are given that

$$a \leq x \leq b$$

we can split the above chained inequality

$$a \leq x \tag{1}$$
$$x \leq b \tag{2}$$

The converse of (1) is

$$x \geq a \tag{3}$$

The converse of (2) is

$$b \geq x \tag{4}$$

Chain (4) and (3)

$$b \geq x \geq a$$

Conversely, we are given that

$$b \geq x \geq a$$

we can split the above chained inequality

$$b \geq x \tag{5}$$
$$x \geq a \tag{6}$$

The converse of (5) is

$$x \leq b \tag{7}$$

The converse of (6) is

$$a \leq x \tag{8}$$

Chain (8) and (7)

$$a \leq x \leq b$$

∎

4 If $a \leq x \leq b$ then $a + c \leq x + c \leq b + c$.
   (Addition of a Constant)

Proof We are given that

$$a \leq x \leq b$$

we can split the above chained inequality

$$a \leq x \tag{1}$$
$$x \leq b \tag{2}$$

add c to both sides of (1)

$$a + c \leq x + c \tag{3}$$

add c to both sides of (2)

$$x + c \leq b + c \tag{4}$$

chain (3) and (4)

$$a + c \leq x + c \leq b + c$$

∎

5 If $c \geq 0$ and $a \leq x \leq b$ then $ac \leq xc \leq bc$.
   (Multiplying by a Nonnegative Constant)

Proof We are given that

$$a \leq x \leq b$$

we can split the above chained inequality

$$a \leq x \tag{1}$$
$$x \leq b \tag{2}$$

multiply both sides of (1) by c

$$ac \leq xc \tag{3}$$

multiply both sides of (2) by c

$$xc \leq bc \tag{4}$$

chain (3) and (4)

$$ac \leq xc \leq bc$$

∎

6 If $a \leq x \leq b$ and $c \leq y \leq d$ then $a + c \leq x + y \leq b + d$. (Addition of Two Chained Inequalities)

Proof We are given that

$$a \leq x \leq b$$

we can split the above chained inequality

$$a \leq x \tag{1}$$
$$x \leq b \tag{2}$$

We are also given that

$$c \leq y \leq d$$

we can split the above chained inequality

$$c \leq y \tag{3}$$
$$y \leq d \tag{4}$$

Add (1) and (3)

$$a + c \leq x + y \tag{5}$$

Add (2) and (4)

$$x + y \leq b + d \tag{6}$$

chain (5) and (6)

$$a + c \leq x + y \leq b + d$$

∎

7 If $a \leq x \leq b$ and $c \leq y \leq d$ then $a - d \leq x - y \leq b - c$. (Subtraction of Two Chained Inequalities)

Proof We are given that

$$a \leq x \leq b \tag{1}$$
$$c \leq y \leq d \tag{2}$$

3.3 Non-Strict Chained Inequality

Multiply through (2) by -1

$$-c \geq -y \geq -d$$
$$-d \leq -y \leq -c \qquad (3)$$

Add (1) and (3)

$$a - d \leq x - y \leq b - c$$

∎

8. Let $a, c \geq 0$. If $a \leq x \leq b$ and $c \leq y \leq d$ then $ac \leq xy \leq bd$. (Multiplication of Two Chained Inequalities)

Proof We are given that

$$a \leq x \leq b$$

we can split the above chained inequality

$$a \leq x \qquad (1)$$
$$x \leq b \qquad (2)$$

We are also given that

$$c \leq y \leq d$$

we can split the above chained inequality

$$c \leq y \qquad (3)$$
$$y \leq d \qquad (4)$$

Multiply (1) and (3)

$$ac \leq xy \qquad (5)$$

Multiply (2) and (4)

$$xy \leq bd \qquad (6)$$

chain (5) and (6)

$$ac \leq xy \leq bd$$

∎

9 Let $a, c > 0$. If $a \leq x \leq b$ and $c \leq y \leq d$ then $\dfrac{a}{d} \leq \dfrac{x}{y} \leq \dfrac{b}{c}$.

(Division of Two Chained Inequalities)

Proof We are given that

$$a \leq x \leq b$$
$$c \leq y \leq d$$

we can split each of the above chained inequality

$$a \leq x \tag{1}$$
$$x \leq b \tag{2}$$
$$c \leq y \tag{3}$$
$$y \leq d \tag{4}$$

Since $0 < c$ and $c \leq y$, by transitivity, $0 < y$. Hence $c, y > 0$. Divide (1) by (4)

$$\frac{a}{d} \leq \frac{x}{y}$$

Divide (2) by (3)

$$\frac{x}{y} \leq \frac{b}{c}$$

Chain the last two inequalities together

$$\frac{a}{d} \leq \frac{x}{y} \leq \frac{b}{c}$$

∎

Exercises 3.3

Let a, b, c, d, x, y be real numbers. Let $a_1, \ldots, a_n, b_1, \ldots, b_n$ and $x_1, \ldots, x_n$ be real numbers. Prove the following.

1   **a** If $a \geq x \geq b$ then $a + c \geq x + c \geq b + c$.
       (Addition of a Constant)

   b If $c \geq 0$ and $a \geq x \geq b$ then $ac \geq xc \geq bc$.
       (Multiplying by a Nonnegative Constant)

c If $a \geq x \geq b$ and $c \geq y \geq d$ then $a + c \geq x + y \geq b + d$.
(Addition of Two Chained Inequalities)

d If $a \geq x \geq b$ and $c \geq y \geq d$ then $a - d \geq x - y \geq b - c$.
(Subtraction of Two Chained Inequalities)

e Let $b, d \geq 0$. If $a \geq x \geq b$ and $c \geq y \geq d$ then $ac \geq xy \geq bd$.
(Multiplication of Two Chained Inequalities)

f Let $b, d > 0$. If $a \geq x \geq b$ and $c \geq y \geq d$ then $\dfrac{a}{d} \geq \dfrac{x}{y} \geq \dfrac{b}{c}$.
(Division of Two Chained Inequalities)

2 a If $c \geq 0$ and $a \leq x \leq b$ then $a - c \leq x \leq b + c$.
(Superset of a Chained Inequality)

b If $c < 0$ and $a \leq x \leq b$ then $ac \geq xc \geq bc$.
(Multiplying by a Negative Constant)

c If
$$a_1 \leq x_1 \leq b_1, \ a_2 \leq x_2 \leq b_2, \ldots, a_n \leq x_n \leq b_n$$
then
$$a_1 + \cdots + a_n \leq x_1 + \cdots + x_n \leq b_1 + \cdots + b_n.$$
(Addition of Arbitrary Number of Chained Inequalities)

d Let $a_1, \ldots, a_n \geq 0$. If
$$a_1 \leq x_1 \leq b_1, \ a_2 \leq x_2 \leq b_2, \ldots, a_n \leq x_n \leq b_n$$
then
$$a_1 \cdots a_n \leq x_1 \cdots x_n \leq b_1 \cdots b_n.$$
(Multiplication of Arbitrary Number of Chained Inequalities)

e If $a \geq 0$ and $a \leq x \leq b$ then $a^n \leq x^n \leq b^n$.
(Power of a Chained Inequality)

3 a If $c \geq 0$ and $a \geq x \geq b$ then $a + c \geq x \geq b - c$.
(Superset of a Chained Inequality)

b If $c < 0$ and $a \geq x \geq b$ then $ac \leq xc \leq bc$.
(Multiplying by a Negative Constant)

c If
$$a_1 \geq x_1 \geq b_1, a_2 \geq x_2 \geq b_2, \ldots, a_n \geq x_n \geq b_n$$
then
$$a_1 + \cdots + a_n \geq x_1 + \cdots + x_n \geq b_1 + \cdots + b_n.$$
(Addition of Arbitrary Number of Chained Inequalities)

d Let $b_1, \ldots, b_n \geq 0$. If
$$a_1 \geq x_1 \geq b_1, a_2 \geq x_2 \geq b_2, \ldots, a_n \geq x_n \geq b_n$$
then
$$a_1 \cdots a_n \geq x_1 \cdots x_n \geq b_1 \cdots b_n.$$
(Multiplication of Arbitrary Number of Chained Inequalities)

e If $b \geq 0$ and $a \geq x \geq b$ then $a^n \geq x^n \geq b^n$.
(Power of a Chained Inequality)

3.4 Mixed Chained Inequalities

1 In this section, we define mixed chained inequalities and prove some of their properties.

2 **Definition of mixed chained inequality.** An inequality such as $1 < 2 \leq 5$ is called a mixed chained inequality. When a chained inequality contains both strict and non-strict inequalities it is called a *mixed chained inequality*.

3 **Properties of mixed chained inequalities.** Here, we prove some properties of mixed chained inequalities, which are summarized in the table below.

name	property
addition of mixed chained inequalities	If $a \leq x < b$ and $c < y \leq d$ then $a + c < x + y < b + d$
subtraction of mixed chained inequalities	If $a \leq x < b$ and $c \leq y < d$ then $a - d < x - y < b - c$
multiplication of mixed chained inequalities	Let $a, c > 0$. If $a \leq x < b$ and $c < y \leq d$ then $ac < xy < bd$
division of mixed chained inequalities	Let $a, c > 0$. If $a \leq x < b$ and $c \leq y < d$ then $\frac{a}{d} < \frac{x}{y} < \frac{b}{c}$.

3.4 Mixed Chained Inequalities

4. Let a, b, c, d, x, y be real numbers.

5. If $a \leq x < b$ and $c < y \leq d$ then $a + c < x + y < b + d$.

    (Addition of Mixed Chained Inequalities)

    Proof We are given that

    $$a \leq x < b$$

    we can split the above chained inequality

    $$a \leq x \qquad (1)$$
    $$x < b \qquad (2)$$

    We are also given that

    $$c < y \leq d$$

    we can split the above chained inequality

    $$c < y \qquad (3)$$
    $$y \leq d \qquad (4)$$

    Add (1) and (3)

    $$a + c < x + y \qquad (5)$$

    Add (2) and (4)

    $$x + y < b + d \qquad (6)$$

    chain (5) and (6)

    $$a + c < x + y < b + d$$

    ■

6. If $a \leq x < b$ and $c \leq y < d$ then $a - d < x - y < b - c$.

    (Subtraction of Mixed Chained Inequalities)

Proof We are given that

$$a \leq x < b \quad (1)$$
$$c \leq y < d \quad (2)$$

Multiply through (2) by -1

$$-c \geq -y > -d$$
$$-d < -y \leq -c \quad (3)$$

Add (1) and (3)

$$a - d < x - y < b - c$$

∎

7   Let $a, c > 0$. If $a \leq x < b$ and $c < y \leq d$ then $ac < xy < bd$. (Multiplication of Mixed Chained Inequalities)

Proof We are given that

$$a \leq x < b$$

we can split the above chained inequality

$$a \leq x \quad (1)$$
$$x < b \quad (2)$$

We are also given that

$$c < y \leq d$$

we can split the above chained inequality

$$c < y \quad (3)$$
$$y \leq d \quad (4)$$

Multiply (1) and (3)

$$ac < xy \quad (5)$$

Multiply (2) and (4)

$$xy < bd \quad (6)$$

chain (5) and (6)

$$ac < xy < bd$$

∎

8   Let $a, c > 0$. If $a \leq x < b$ and $c \leq y < d$ then $\dfrac{a}{d} < \dfrac{x}{y} < \dfrac{b}{c}$.

(Division of Mixed Chained Inequalities)

Proof We are given that

$$a \leq x < b$$
$$c \leq y < d$$

we can split each of the above chained inequality

$$a \leq x \qquad (1)$$
$$x < b \qquad (2)$$
$$c \leq y \qquad (3)$$
$$y < d \qquad (4)$$

Since $0 < c$ and $c \leq y$, by transitivity, $0 < y$. Hence $c, y > 0$. Divide (1) by (4)

$$\frac{a}{d} < \frac{x}{y}$$

Divide (2) by (3)

$$\frac{x}{y} < \frac{b}{c}$$

Chain the last two inequalities together

$$\frac{a}{d} < \frac{x}{y} < \frac{c}{d}$$

∎

Exercises 3.4

1 Prove the following for all real numbers a, b, x.

   a If $a < x < b$ then $a \leq x < b$.

   b If $a < x < b$ then $a < x \leq b$.

   c If $a < x < b$ then $a \leq x \leq b$.

   d If $a \leq x < b$ then $a \leq x \leq b$.

56   Chapter 3   Chained Inequality

   e   If $a < x \leq b$ then $a \leq x \leq b$.

2   Let a, b, c, d, x, y be real numbers. Prove the following.

   a $a \leq x \leq b$ implies $a \leq x \leq y$ or $y \leq x \leq b$.

   b $a < x < b$ implies $a < x < y$ or $y \leq x < b$.

   c $a \leq x < b$ if and only if $b > x \geq a$.
       (Converse of Mixed Chained Inequality I)

   d $a < x \leq b$ if and only if $b \geq x > a$.
       (Converse of Mixed Chained Inequality II)

   e   If $c \geq 0$ and $a \leq x < b$ then $a - c \leq x < b + c$.
       (Superset of Chained Inequalities)

   f   If $a \leq x < b$ then $a + c \leq x + c < b + c$.
       (Addition of a Constant)

   g   If $c > 0$ and $a \leq x < b$ then $ac \leq xc < bc$.
       (Multiplying by a Positive Constant)

   h   If $c < 0$ and $a \leq x < b$ then $ac \geq xc > bc$.
       (Multiplying by a Negative Constant)

   i   If $a \leq x < b$ and $c \leq y < d$ then $a + c \leq x + y \leq b + d$.
       (Addition of Mixed Chained Inequalities I)

   j   If $a \leq x < b$ and $c < y \leq d$ then $a + c < x + y < b + d$.
       (Addition of Mixed Chained Inequalities II)

   k   If $a \leq x < b$ and $c < y \leq d$ then $a - d \leq x - y \leq b - c$.
       (Subtraction of Mixed Chained Inequalities I)

   l   If $a \leq x < b$ and $c \leq y < d$ then $a - d < x - y < b - c$.
       (Subtraction of Mixed Chained Inequalities II)

   m   Let $a, c > 0$. If $a \leq x < b$ and $c \leq y < d$ then $ac \leq xy < bd$.
       (Multiplication of Mixed Chained Inequalities I)

   n   Let $a, c > 0$. If $a \leq x < b$ and $c < y \leq d$ then $ac < xy < bd$.
       (Multiplication of Mixed Chained Inequalities II)

o   Let $a, c > 0$. If $a \leq x < b$ and $c < y \leq d$ then $\dfrac{a}{d} \leq \dfrac{x}{y} \leq \dfrac{b}{c}$.
   (Division of Mixed Chained Inequalities I)

p   Let $a, c > 0$. If $a \leq x < b$ and $c \leq y < d$ then $\dfrac{a}{d} < \dfrac{x}{y} < \dfrac{b}{c}$.
   (Division of Mixed Chained Inequalities II)

q   If $a \geq 0$ and $a \leq x < b$ then $a^n \leq x^n < b^n$.
   (Power of Mixed Chained Inequality)

3.5 Extended Chained Inequality

1 In this section, we extend the definition of chained inequality to allow for chaining more that two inequalities together.

2 **Definition of chained inequality.** A short way of writing $1 < 2$, $2 < 3$ and $3 < 4$ is
$$1 < 2 < 3 < 4.$$
Similarly, we may write $5 > 4$, $4 \geq 2$, $2 > 1$ and $1 > 0$ as
$$5 > 4 \geq 2 > 1 > 0.$$

In general, we may use this notation to write any number of inequalities in a chain provided they can be arranged such that each inequality shares a number with the next inequality.

3 **Examples of chained inequalities.** Here are some examples of chained inequalities: we write

   • $w < x \leq y < z$ if and only if $w < x$, $x \leq y$ and $y < z$.
   • $v \geq w \geq x \geq y \geq z$ if and only if $v \geq w$, $w \geq x$, $x \geq y$ and $y \geq z$.

An inequality such as one of the types above is called a *chained inequality*.

3.6 An Inequality Due to Archimedes

1 In this section, we prove that for all integers n greater than 1,
$$1^2 + 2^2 + \cdots + (n-1)^2 < \dfrac{n^3}{3} < 1^2 + 2^2 + \cdots + n^2.$$

Before proving this inequality, we prove some lemmas and corollaries.

2   Let n be a positive integer.

3   $n(2n - 1) < 2n^2 < (n + 1)(2n + 1)$

Proof

$$-2 < -1 < 1$$
$$0 \leq 0 < 1$$

Add n to each chained inequality

$$n - 2 < n - 1 < n + 1$$
$$n \leq n < n + 1$$

Multiply both chained inequalities together

$$n(n - 2) < n(n - 1) < (n + 1)^2$$

Add $n(n + 1)$ to both sides

$$n(n - 2) + n(n + 1) < n(n - 1) + n(n + 1) < (n + 1)^2 + n(n + 1)$$
$$n(n - 2 + n + 1) < n(n - 1 + n + 1) < (n + 1)(n + 1 + n)$$
$$n(2n - 1) < n(2n) < (n + 1)(2n + 1)$$
$$n(2n - 1) < 2n^2 < (n + 1)(2n + 1)$$

■

4   $(n + 1)(2n + 1) > 2n^2$

Proof   From a previous result $n(2n - 1) < 2n^2 < (n + 1)(2n + 1)$. Thus $2n^2 < (n + 1)(2n + 1)$. Hence, $(n + 1)(2n + 1) > 2n^2$.   ■

5   $(n - 1)(2n - 1) < 2n^2$

Proof   We know that $n - 1 < n$. Multiply both sides by $2n - 1$. Hence $(n - 1)(2n - 1) < n(2n - 1)$. From a previous result

$$n(2n - 1) < 2n^2$$

Since $(n - 1)(2n - 1) < n(2n - 1)$ we can write

$$(n - 1)(2n - 1) < 2n^2$$

■

6 $(n-1)(2n-1) < 2n^2 < (n+1)(2n+1)$

Proof I From previous results we get $(n-1)(2n-1) < 2n^2$ and $2n^2 < (n+1)(2n+1)$. Chaining both inequalities, we obtain $(n-1)(2n-1) < 2n^2 < (n+1)(2n+1)$. ∎

7 $1^2 + 2^2 + \cdots + (n-1)^2 < \dfrac{n^3}{3} < 1^2 + 2^2 + \cdots + n^2$

Proof

From a previous result
$$(n-1)(2n-1) < 2n^2 < (n+1)(2n+1)$$
Since n is positive, we can multiply the previous inequality by n.
$$n(n-1)(2n-1) < 2n^3 < n(n+1)(2n+1)$$
Divide by 6
$$\frac{(n-1)n(2n-1)}{6} < \frac{2n^3}{6} < \frac{n(n+1)(2n+1)}{6} \qquad (1)$$
We know that
$$1^2 + 2^2 + \cdots + n^2 = \frac{n(n+1)(2n+1)}{6}$$
And
$$1^2 + 2^2 + \cdots + (n-1)^2 = \frac{(n-1)(n-1+1)(2(n-1)+1)}{6}$$
$$= \frac{(n-1)n(2n-1)}{6}$$
Thus we can write (1) as
$$1^2 + 2^2 + \cdots + (n-1)^2 < \frac{n^3}{3} < 1^2 + 2^2 + \cdots + n^2$$
∎

Exercises 3.6

Let n be an integer greater than 1. Prove the following

1 $(n+1)(n+2) > n^2 + 1$

2 $(n-1)(n-2) < n^2 + 1$

3 $(n-1)(n-2) < n^2 + 1 < (n+1)(n+2)$

CHAPTER 4

Absolute Value

4.1 Absolute Value

1. In this section, we define absolute value and its notation. Then we give examples and prove some properties of absolute value.

2. For all real numbers x, the absolute value of x is denoted by

$$|x|$$

(Absolute Value Notation)

3. For all real numbers x, the absolute value of x is defined thus

   1. $|x| = x$ for all nonnegative real numbers
   2. $|x| = -x$ for all negative real numbers

   (Definition of Absolute Value)

4. Let us find the absolute value of the numbers below.

5. 5.

   Solution Since $5 \geq 0$, it follows that $|5| = 5$.

6. -5.

   Solution Since $-5 < 0$, it follows that $|-5| = -(-5)$. Hence $|-5| = 5$

7. 0.

   Solution Since $0 \geq 0$, it follows that $|0| = 0$.

8 **Properties of absolute value.** Here, we provide an alternative definition of absolute value and prove the following properties:

- $|a| = |-a|$
- $|a| = \pm a$
- $|a| = 0$ if and only if $a = 0$
- $||a|| = |a|$
- $|a| \geq a$
- $|a| \geq -a$
- $|a| + a \geq 0$
- $|a| - a \geq 0$
- $|a| \geq 0$

9 Let a be a real number.

10 An alternative definition of the absolute value function is:

1. $|a| = a$ for all $a > 0$
2. $|a| = -a$ for all $a \leq 0$

Proof When $a > 0$, it is true that $a \geq 0$. Hence the definition of absolute value gives $|a| = a$.

When $a < 0$, the definition of absolute value gives $|a| = -a$. Finally, when $a = 0$ we have that $|a| = 0 = -0$. Hence $|a| = -a$. If we combine the case $a < 0$ and $a = 0$, we have that when $a \leq 0$, $|a| = -a$. ■

11 $|a| = |-a|$

Proof When $a \geq 0$, $|a| = a$. Also $-a \leq 0$, hence $|-a| = -(-a)$. Thus $|-a| = a$. Consequently, $|a| = |-a|$.

When $a < 0$, $|a| = -a$. Also $-a > 0$, hence $|-a| = -a$. Consequently, $|a| = |-a|$.

In both cases, $|a| = |-a|$. ■

12 $|a| = \pm a$

Proof When $a \geq 0$ the definition of the absolute value function gives $|a| = a$. Hence $|a| = \pm a$. When $a < 0$ the definition of the absolute value function gives $|a| = -a$. Hence $|a| = \pm a$. Therefore, in both cases, $|a| = \pm a$. ■

4.1 Absolute Value

13 $a = \pm|a|$

Proof From a previous result, $|a| = \pm a$. Thus, $|a| = a$ or $|a| = -a$. If $|a| = a$ then $\pm|a| = a$. If $|a| = -a$ then $-|a| = a$. Hence $\pm|a| = a$. In both cases, $\pm|a| = a$. ∎

14 $|a| = 0$ if and only if $a = 0$

Proof We are given that $|a| = 0$. We know that $a = \pm|a|$. Thus $a = \pm 0$. Therefore $a = 0$. Conversely, suppose $a = 0$. Since $a \geq 0$, the definition of absolute value function gives $|a| = a$. Therefore $|a| = 0$. ∎

15 $||a|| = |a|$

Proof Let $z = |a|$. We know that $z \geq 0$. Hence the definition of the absolute value function gives $|z| = z$. Hence $||a|| = |a|$. ∎

16 $|a| \geq a$

Proof It is either $a \geq 0$ or $a < 0$. Consider the case where $a \geq 0$.

$$|a| = a$$

Thus,

$$|a| \geq a$$

Consider the case where $a < 0$.

$$a < 0 \tag{1}$$

multiply both sides by -1

$$-a > 0 \tag{2}$$

apply transitivity to (1) and (2)

$$-a > a \tag{3}$$

Since $a < 0$

$$|a| = -a$$

Hence we can substitute for $-a$ in (3)

$$|a| > a$$
$$|a| \geq a$$

In both cases we have that

$$|a| \geq a$$

∎

17 $|a| \geq -a$

Proof By applying a previous lemma, $|-a| \geq -a$. We know that $|a| = |-a|$. Therefore by substituting for $|-a|$ in the previous inequality we obtain $|a| \geq -a$. ∎

18 $|a| + a \geq 0$

Proof We know that $|a| \geq -a$. Adding a to both sides we obtain $|a|+a \geq 0$. ∎

19 $|a| - a \geq 0$

Proof We know that $|a| \geq a$. Subtracting a from both sides we obtain $|a| - a \geq 0$. ∎

20 $|a| \geq 0$

Proof We know that $|a| \geq a$ and $|a| \geq -a$. Adding both inequalities gives $2|a| \geq 0$. Dividing both sides by 2, we obtain $|a| \geq 0$. ∎

4.2 Absolute Value and Squares

1 In this section, we prove some theorems that show the relationship between absolute values and squares:

- If $|w| < |x|$ then $w^2 < x^2$.
- If $|w| > |x|$ then $w^2 > x^2$.
- $|w| = |x|$ if and only if $w^2 = x^2$.

2 Let a, w, x be real numbers.

3 $|a|^2 = a^2$

Proof

$$|a| = \pm a$$

Square both sides

$$\begin{aligned}|a|^2 &= (\pm a)^2 \\ &= (\pm 1)^2 a^2 \\ &= 1 \cdot a^2 \\ &= a^2\end{aligned}$$

∎

4.2 Absolute Value and Squares

4. If $|w| < |x|$ then $w^2 < x^2$.

   Proof We are given that $|w| < |x|$. Since absolute values are nonnegative, we can square both sides of the inequality to obtain $|w|^2 < |x|^2$. Substituting $|w|^2 = w^2$ and $|x|^2 = x^2$ in the inequality we get $w^2 < x^2$. ∎

5. If $|w| > |x|$ then $w^2 > x^2$.

   Proof We are given that $|w| > |x|$. Since absolute values are nonnegative, we can square both sides of the inequality to obtain $|w|^2 > |x|^2$. Substituting $|w|^2 = w^2$ and $|x|^2 = x^2$ in the inequality we get $w^2 > x^2$. ∎

6. If $|w| \neq |x|$ then $w^2 \neq x^2$.

   Proof We are given that
   $$|w| \neq |x|$$
   Thus
   $$|w| < |x| \quad \text{or} \quad |w| > |x|$$
   From No.1 and No.2 we have
   $$w^2 < x^2 \quad \text{or} \quad w^2 > x^2$$
   Hence
   $$w^2 \neq x^2$$
   ∎

7. If $w^2 = x^2$ then $|w| = |x|$.

   Proof This is the contrapositive of the previous result. ∎

8. If $|w| = |x|$ then $w^2 = x^2$.

   Proof We are given that
   $$|w| = |x|$$
   Square both sides
   $$|w|^2 = |x|^2$$
   $$w^2 = x^2$$
   ∎

9. $|w| = |x|$ if and only if $w^2 = x^2$.

   Proof This is the conjunction of the last two results. ∎

4.3 Absolute Value and Powers

1. In this section, we prove the following theorems.

   - $|x^n| = |x|^n$
   - $|w| = |x|$ if and only if $|w^n| = |x^n|$.
   - If $w, x \geq 0$ and $w^n > x^n$ then $w > x$.

2. Let w, x be real numbers and let n be a positive integer.

3. If $x \geq 0$ then $|x^n| = |x|^n$

   Proof We are given that

   $$x \geq 0 \qquad (1)$$

   Thus

   $$|x| = x \qquad (2)$$

   Raise both sides of (1) to the power n

   $$x^n \geq 0^n$$
   $$x^n \geq 0$$
   $$|x^n| = x^n$$

   From (2) we can substitute for x the RHS

   $$|x^n| = |x|^n$$

   ∎

4. If $x < 0$ then $|x^n| = |x|^n$

   Proof We are given that

   $$x < 0$$
   $$-x > 0$$

   Due to the previous result, we can write

   $$|-x|^n = |(-x)^n|$$

4.3 Absolute Value and Powers

$$|x|^n = |(-1)^n x^n|$$
$$= |(-1)^n||x^n|$$
$$= |\pm 1||x^n|$$
$$= 1 \cdot |x^n|$$
$$= |x^n|$$

■

5   $|x^n| = |x|^n$

Proof   It is either $x \geq 0$ or $x < 0$. When $x \geq 0$ we have that $|x^n| = |x|^n$. When $x < 0$ we have that $|x^n| = |x|^n$. Hence in both cases $|x^n| = |x|^n$.   ■

6   If $|w| < |x|$ then $|w^n| < |x^n|$.

Proof   We are given that $|w| < |x|$. Since absolute values are nonnegative, we can raise both sides of the inequality to the power n. Thus $|w|^n < |x|^n$. Since $|w|^n = |w^n|$ and $|x|^n = |x^n|$, we can write the previous inequality as $|w^n| < |x^n|$.   ■

7   If $|w| > |x|$ then $|w^n| > |x^n|$.

Note   The proof is similar to the proof of the previous result. Do it as an exercise.

8   If $|w| \neq |x|$ then $|w^n| \neq |x^n|$.

Proof   We are given that

$$|w| \neq |x|$$

Thus

$$|w| < |x| \quad \text{or} \quad |w| > |x|$$

From No.4 and No.5 we have

$$|w^n| < |x^n| \quad \text{or} \quad |w^n| > |x^n|$$

Hence

$$|w^n| \neq |x^n|$$

■

9 If $|w^n| = |x^n|$ then $|w| = |x|$.

Proof This is the contrapositive of No.6. ∎

10 If $|w| = |x|$ then $|w^n| = |x^n|$.

Proof We are given that $|w| = |x|$. Raise both sides to the power n. Thus $|w|^n = |x|^n$. Hence $|w^n| = |x^n|$. ∎

11 $|w| = |x|$ if and only if $|w^n| = |x^n|$.

Proof This is the conjunction of No.7 and No.8. ∎

12 $|w| = |x|$ if and only if $|w|^n = |x|^n$.

Proof We know that $|x|^n = |x^n|$ and $|w|^n = |w^n|$. It follows that $|w|^n = |x|^n$ if and only if $|w^n| = |x^n|$. From a previous result, we know that $|w^n| = |x^n|$ if and only if $|w| = |x|$. Thus, $|w^n| = |x^n|$ if and only if $|w| = |x|$. In other words, $|w| = |x|$ if and only if $|w|^n = |x|^n$. ∎

13 If $w, x \geq 0$ and $w^n \geq x^n$ then $w \geq x$.

Proof Assume $w < x$. Since $w \geq 0$, we can raise both sides of the inequality by the nth power. Thus $w^n < x^n$. Hence $w^n < x^n$ and $w^n \geq x^n$. This contradiction implies that the statement $w < x$ is wrong. Hence $w \geq x$. ∎

14 If $w, x \geq 0$ and $w^n > x^n$ then $w \neq x$.

Proof Assume $w = x$. Thus $w^n = x^n$. Hence $w^n \leq x^n$. This contradicts with the hypothesis that $w^n > x^n$. Thus the assumption that $w = x$ must be wrong. Therefore $w \neq x$. ∎

15 If $w, x \geq 0$ and $w^n > x^n$ then $w > x$.

Proof We are given that $w, x \geq 0$ and $w^n > x^n$. Thus $w, x \geq 0$ and $w^n \geq x^n$. Hence $w \geq x$.

From A previous result we obtain $w \neq x$. Since $w \geq x$ and $w \neq x$, it follows that $w > x$. ∎

Exercises 4.3

Let w, x be real numbers. Prove the following.

1 If $|w| \geq |x|$ then $|w^n| \geq |x^n|$.

2 If $|w| \leq |x|$ then $|w^n| \leq |x^n|$.

3 If $|w| \geq |x|$ then $|w|^n \geq |x|^n$.

4 If $|w| \leq |x|$ then $|w|^n \leq |x|^n$.

5 If $|w| > |x|$ then $|w|^n > |x|^n$.

6 If $|w| < |x|$ then $|w|^n < |x|^n$.

7 If $w, x \geq 0$ and $w^n < x^n$ then $w < x$.

4.4 Absolute Value and Triangle Inequality

1 In this section, we prove the triangle inequality among other theorems:

- $|ab| = |a||b|$
- $|a| = |b|$ if and only if $a = \pm b$.
- If $b \neq 0$ then $\left|\dfrac{a}{b}\right| = \dfrac{|a|}{|b|}$.
- $|a| \leq b$ if and only if $-b \leq a \leq b$.
- $|a| < b$ if and only if $-b < a < b$.
- $|a + b| \leq |a| + |b|$ (Triangle inequality)
- $|a - b| \geq |a| - |b|$
- $|a - b| \geq \big||a| - |b|\big|$
- $|a - b| \leq |a - c| + |c - b|$

2 Let a, b, c be real numbers.

3 $|a - b| = 0$ if and only if $a = b$.

Proof We know that $|a - b| = 0$ if and only if $a - b = 0$. Also, $a - b = 0$ if and only if $a = b$. Consequently, $|a - b| = 0$ if and only if $a = b$. ∎

4 $|a| + |b| \geq a + b$

Proof We know that $|a| \geq a$ and $|b| \geq b$. Adding both inequalities, we obtain $|a| + |b| \geq a + b$. ∎

5. $|ab| = |a||b|$

Proof It is either a and b are both nonnegative or one of them is negative. Consider the case where a and b are both nonnegative. Thus,

$$a \geq 0 \qquad (1)$$
$$b \geq 0 \qquad (2)$$

The definition of absolute value function gives

$$|a| = a$$
$$|b| = b$$

Multiply both equations together

$$|a||b| = ab \qquad (3)$$

Multiply inequalities (1) and (2) together

$$ab \geq 0$$

The definition of absolute value function gives

$$|ab| = ab$$

Use (3) to substitute for ab in the above equation

$$|ab| = |a||b|$$

Consider the case where one of a, b is negative. Thus either $a \geq 0$ or $-a \geq 0$. Hence $\pm a \geq 0$. Similarly $\pm b \geq 0$.

$$\pm a \geq 0 \qquad (4)$$
$$\pm b \geq 0 \qquad (5)$$

From (4) and (5) the previous case gives

$$|(\pm a)(\pm b)| = |\pm a||\pm b|$$
$$|\pm ab| = |a||b|$$
$$|ab| = |a||b|$$

In all cases $|ab| = |a||b|$. ∎

6 $|a| = |b|$ if and only if $a = \pm b$.

Proof First, we prove that if $|a| = |b|$ then $a = \pm b$.
$$|a| = |b|$$
$$\pm a = \pm b$$
$$a = \pm b$$

Next, we prove that if $a = \pm b$ then $|a| = |b|$.
$$a = \pm b$$

Find absolute value of both sides
$$|a| = |\pm b|$$
$$= |\pm 1| \cdot |b|$$
$$= 1 \cdot |b|$$
$$= |b|$$

■

7 If $a \neq 0$ then $\left|\dfrac{1}{a}\right| = \dfrac{1}{|a|}$.

Proof Since $a \neq 0$, $1/a$ is a real number and $|a| > 0$. Hence a and $1/a$ are real numbers. By a previous lemma we get $|a \cdot (1/a)| = |a||1/a|$. Thus $|1| = |a||1/a|$. It follows that $1 = |a||1/a|$. Dividing both sides by $|a|$ gives
$$\frac{1}{|a|} = \left|\frac{1}{a}\right|$$

■

8 If $b \neq 0$ then $\left|\dfrac{a}{b}\right| = \dfrac{|a|}{|b|}$.

Proof Since $b \neq 0$ then $1/b$ is a real number. Hence a and $1/b$ are real numbers. By a previous lemma we get
$$\left|a \cdot \left(\frac{1}{b}\right)\right| = |a| \cdot \left|\frac{1}{b}\right|$$
$$\left|\frac{a}{b}\right| = |a| \cdot \frac{1}{|b|}$$
$$= \frac{|a|}{|b|}$$

■

Chapter 4 Absolute Value

9   $|a| \leq b$ if and only if $-b \leq a \leq b$.

Proof First, we prove that if $|a| \leq b$ then $-b \leq a \leq b$.

$$b \geq |a| \tag{1}$$

From a previous lemma

$$|a| \geq a \tag{2}$$
$$|a| \geq -a \tag{3}$$

Apply transitivity to (1) and (2)

$$b \geq a \tag{4}$$

Apply transitivity to (1) and (3)

$$b \geq -a$$

Multiply both sides by -1

$$-b \leq a \tag{5}$$

Chain (4) and (5)

$$-b \leq a \leq b$$

Next, we prove that if $-b \leq a \leq b$ then $|a| \leq b$.

$$-b \leq a \leq b$$

Thus

$$a \leq b \tag{1}$$
$$-b \leq a$$

Multiply both sides of the previous inequality by -1

$$b \geq -a \tag{2}$$

From (1) and (2) we have that $b \geq a$ and $b \geq -a$. Since one of a and $-a$ is $|a|$, it follows that $b \geq |a|$. ∎

10   $|a| < b$ if and only if $-b < a < b$.

Proof We are given that $|a| < b$. Thus $|a| \leq b$. Hence the previous theorem gives $-b \leq a \leq b$. Since $b > |a|$, it follows that $b > a$ and $b > -a$. Hence $b \neq a$ and $b \neq -a$. Thus $b \neq a$ and $-b \neq a$. Hence we may write the previous chained inequality as $-b < a < b$.

Conversely, given $-b < a < b$ then $-b \leq a \leq b$. Hence the previous theorem gives $|a| \leq b$. From the strict chained inequality we have that $b \neq a$ and $-b \neq a$. Hence $b \neq a$ and $b \neq -a$. It follows that $|a| \neq b$. Therefore we may write $|a| < b$. ∎

11 $|a + b| \leq |a| + |b|$

(Triangle Inequality)

Proof Let

$$z = |a| + |b|$$

We know that

$$a \leq |a| \quad (1)$$
$$b \leq |b| \quad (2)$$

Also,

$$-a \leq |a| \quad (3)$$
$$-b \leq |b| \quad (4)$$

Add (1) and (2)

$$a + b \leq |a| + |b|$$
$$a + b \leq z \quad (5)$$

Add (3) and (4)

$$-a - b \leq |a| + |b|$$
$$-a - b \leq z$$

Multiply both sides by -1

$$a + b \geq -z \quad (6)$$

Chain (5) and (6)

$$-z \leq a + b \leq z \quad (7)$$

Thus
$$|a + b| \le z$$
$$|a + b| \le |a| + |b|$$

∎

12 $|a - b| \ge |a| - |b|$

Proof For all real numbers x, y the triangle inequality gives
$$|x + y| \le |x| + |y|$$
Put $x = a - b$ and $y = b$
$$|(a - b) + b| \le |a - b| + |b|$$
$$|a| \le |a - b| + |b|$$
$$|a| - |b| \le |a - b|$$
$$|a - b| \ge |a| - |b|$$

∎

13 $|a - b| \ge |b| - |a|$

Proof We know that
$$|b - a| \ge |b| - |a|$$
$$|-(b - a)| \ge |b| - |a|$$
$$|a - b| \ge |b| - |a|$$

∎

14 $|a - b| \ge \big||a| - |b|\big|$

Proof Let
$$z = |a| - |b|$$
We know that
$$|a - b| \ge z \tag{1}$$

and
$$|a - b| \geq -z$$

multiply both sides by -1

$$-|a - b| \leq z \tag{2}$$

Chain (1) and (2)

$$-|a - b| \leq z \leq |a - b|$$

The above inequality implies

$$|z| \leq |a - b|$$
$$\bigl||a| - |b|\bigr| \leq |a - b|$$
$$|a - b| \geq \bigl||a| - |b|\bigr|$$

∎

15 $|a - b| \leq |a - c| + |c - b|$

Proof For all real numbers x, y the triangle inequality gives

$$|x + y| \leq |x| + |y|$$

Put $x = a - c$ and $y = c - b$

$$|(a - c) + (c - b)| \leq |a - c| + |c - b|$$
$$|a - c + c - b| \leq |a - c| + |c - b|$$
$$|a - b| \leq |a - c| + |c - b|$$

∎

Exercises 4.4

Let a, b, c be real numbers. Let n be a positive integer and let $a_1, \ldots, a_n$ be real numbers.

1 $|abc| = |a| \cdot |b| \cdot |c|$

2 $|a_1 \cdots a_n| = |a_1| \cdots |a_n|$

3 $|a + b + c| \leq |a| + |b| + |c|$

4 $|a_1 + \cdots + a_n| \leq |a_1| + \cdots + |a_n|$

4.5 Absolute Value and Composite Numbers

1. A composite number is an integer which is a product of two integers greater than 1. Examples are $2 \cdot 3$, $4 \cdot 5$, $3 \cdot 3$ and $7 \cdot 2$.

2. We know that 3×5 is a composite integer. Observe that 3 is between 1 and 3×5. That is, $1 < 3 < 3 \times 5$. Similarly, $1 < 5 < 3 \times 5$.

3. The above property is true for all composite integers. Let a, b be integers greater than 1. Then $1 < a < ab$ and $1 < b < ab$. Before proving this theorem, we prove some lemmas and corollaries.

4. Using a similar method, we prove that for all nonzero integers a, b, if $|a| + |b| \le |ab|$ then $1 < |a| < |ab|$ and $1 < |b| < |ab|$.

5. Let a, b be nonzero integers.

6. $1 \le |a| \le |ab|$

   Proof Observe that

   $$ab \ne 0$$
   $$|ab| > 0$$
   $$|a||b| > 0 \qquad (1)$$

   divide both sides of (1) by $|b|$

   $$|a| > 0$$
   $$|a| \ge 1 \qquad (2)$$

   Now, divide both sides of (1) by $|a|$

   $$|b| > 0$$
   $$|b| \ge 1$$

   multiply both sides by $|a|$

   $$|a||b| \ge |a|$$
   $$|ab| \ge |a| \qquad (3)$$

   From (2) and (3) we can conclude that

   $$1 \le |a| \le |ab|$$

   ∎

4.5 Absolute Value and Composite Numbers

7   If $a, b > 1$ then
$$1 < a < ab \quad \text{and} \quad 1 < b < ab.$$

Proof   Since $a, b > 1$, it follows that $ab > 1$ and $1 \leq |a| \leq |ab|$. Since $a, ab > 1$, we have that $|a| = a$ and $|ab| = ab$. Hence, $1 \leq a \leq ab$. Since $a > 1$, we may write $1 \leq a \leq ab$ as $1 < a \leq ab$. Multiplying $a \geq a$ and $b > 1$, we get $ab > a$. Thus we may write $1 < a \leq ab$ as $1 < a < ab$. By a similar argument, $1 < b < ab$. Thus, $1 < a < ab$ and $1 < b < ab$. ∎

8   If $|a| + |b| \leq |ab|$ then
$$1 < |a| < |ab| \quad \text{and} \quad 1 < |b| < |ab|.$$

Proof   From a previous result,
$$1 \leq |a| \leq |ab|$$
Thus
$$1 \leq |a|$$
$$0 < |a| \tag{1}$$
From a previous result,
$$1 \leq |b| \leq |ab|$$
$$1 \leq |b|$$
$$0 < |b| \tag{2}$$
From the hypothesis
$$|a| + |b| \leq |ab|$$
Thus
$$|a| \leq |ab| - |b| \tag{3}$$
and
$$|b| \leq |ab| - |a| \tag{4}$$
Apply transitivity to (2) and (4)
$$0 < |ab| - |a|$$
$$|a| < |ab| \tag{5}$$
Apply transitivity to (1) and (3)
$$0 < |ab| - |b| \tag{3}$$
$$|b| < |ab|$$

$$|b| < |a||b|$$

divide both sides by $|b|$

$$1 < |a| \tag{6}$$

From (5) and (6) we get

$$1 < |a| < |ab|$$

■

By a similar argument, we get $1 < |b| < |ab|$. Therefore,

$$1 < |a| < |ab| \quad \text{and} \quad 1 < |b| < |ab|.$$

Exercises 4.5

Let a, b be nonzero integers. Prove the following.

1 If $|a| \neq 1$ and $|a| \neq |ab|$ then
$$1 < |a| < |ab|$$

2 If $|a| \neq 1$ and $|b| \neq 1$ then
$$1 < |a| < |ab|$$

3 If $|a| \neq |ab|$ and $|b| \neq |ab|$ then
$$1 < |a| < |ab|$$

4.6 Absolute Value and Chained Inequality

1. In this section, we prove theorems that allows us to convert any chained inequality such as $-6 \leq x \leq 3$ to a simpler inequality such as $|2x + 3| \leq 9$.

    The theorems are listed below. For all real numbers a, b, x,

    1. $a \leq x \leq b$ if and only if $\left|x - \dfrac{a+b}{2}\right| \leq b - \dfrac{a+b}{2}$.
    2. $a \leq x \leq b$ if and only if $|2x - (a+b)| \leq b - a$

2. The concept of arithmetic mean will make proving the above theorems easier. So we define it below.

3 **Definition of arithmetic mean.** The arithmetic mean of two real numbers is defined to be half the sum of the numbers. Let a, b be real numbers. The arithmetic mean of a and b is
$$\frac{a+b}{2}.$$

4 **Absolute value and chained inequality.** Let a, b, x be real numbers and let m be the arithmetic mean of a and b.

5 $a - m = m - b$

Proof
$$\frac{a+b}{2} = m$$

multiply both sides by 2

$$a + b = 2m$$
$$a + b - m = m$$
$$a - m = m - b$$

∎

6
$$a \leq x \leq b$$

if and only if
$$|x - m| \leq b - m.$$

Proof First, we prove that if $a \leq x \leq b$ then $|x - m| \leq b - m$.
$$a \leq x \leq b$$

subtract m throughout

$$a - m \leq x - m \leq b - m$$

Since $a - m = m - b$ we can write

$$m - b \leq x - m \leq b - m$$
$$-(b - m) \leq x - m \leq b - m$$

Thus,
$$|x - m| \leq b - m$$

Next, we prove that if $|x - m| \leq b - m$ then $a \leq x \leq b$.
$$|x - m| \leq b - m$$

Thus,
$$-(b - m) \leq x - m \leq b - m$$
$$m - b \leq x - m \leq b - m$$

Since $a - m = m - b$ we can write
$$a - m \leq x - m \leq b - m$$

add m throughout
$$a \leq x \leq b$$

Thus we can conclude that
$$a \leq x \leq b$$

if and only if
$$|x - m| \leq b - m$$

∎

7 $a \leq x \leq b$ if and only if $\left|x - \dfrac{a+b}{2}\right| \leq b - \dfrac{a+b}{2}$.

Proof We are given that $a \leq x \leq b$. From a previous result,
$$|x - m| \leq b - m,$$
where m is the arithmetic mean of a and b. Since $m = \dfrac{a+b}{2}$, we may write
$$\left|x - \dfrac{a+b}{2}\right| \leq b - \dfrac{a+b}{2}.$$

Conversely, we are given that
$$\left|x - \dfrac{a+b}{2}\right| \leq b - \dfrac{a+b}{2}$$

Let $m = (a + b)/2$. Thus

$$|x - m| \leq b - m$$

From a previous result,

$$a \leq x \leq b$$

∎

8   $a \leq x \leq b$ if and only if $|2x - (a + b)| \leq b - a$

Proof   We are given that $a \leq x \leq b$. Thus,

$$\left|x - \frac{a + b}{2}\right| \leq b - \frac{a + b}{2}$$

Multiply both sides by $|2|$

$$|2| \cdot \left|x - \frac{a + b}{2}\right| \leq |2| \cdot (b - \frac{a + b}{2})$$

$$\left|2 \cdot \left(x - \frac{a + b}{2}\right)\right| \leq 2 \cdot (b - \frac{a + b}{2})$$

$$|2x - (a + b)| \leq 2b - (a + b)$$

$$|2x - (a + b)| \leq b - a$$

Conversely, we are given that $|2x - (a + b)| \leq b - a$. Thus,

$$|2x - (a + b)| \leq 2b - (a + b)$$

$$\left|2 \cdot \left(x - \frac{a + b}{2}\right)\right| \leq 2 \cdot (b - \frac{a + b}{2})$$

$$|2| \cdot \left|x - \frac{a + b}{2}\right| \leq |2| \cdot (b - \frac{a + b}{2})$$

Divide both sides by $|2|$

$$\left|x - \frac{a + b}{2}\right| \leq b - \frac{a + b}{2}$$

Thus,

$$a \leq x \leq b$$

∎

Exercises 4.6

1 Let a, b, x be real numbers. Let $m = \dfrac{a+b}{2}$. Prove the following.

   a $a < x < b$ if and only if $|x - m| < b - m$.

   b $a < x < b$ if and only if $|2x - (a + b)| < b - a$

2 Let a, b, x be real numbers. Let $m = \dfrac{a+b}{2}$. Prove the following.

   a $a \geq x \geq b$ if and only if $|m - x| \geq m - a$.

   b $a \geq x \geq b$ if and only if $|a + b - 2x| \geq b - a$

   c $a > x > b$ if and only if $|m - x| > m - a$.

   d $a > x > b$ if and only if $|a + b - 2x| > b - a$

3 Let a, b, x be real numbers. Let $m = \dfrac{a+b}{2}$. Prove the following.

   a $a \leq x \leq b$ if and only if $m - b \leq |m - x|$.

   b $a \leq x \leq b$ if and only if $a - b \leq |a + b - 2x|$

   c $a < x < b$ if and only if $m - b < |m - x|$.

   d $a < x < b$ if and only if $a - b < |a + b - 2x|$

4 Let a, b, x be real numbers. Let $m = \dfrac{a+b}{2}$. Prove the following.

   a $a \geq x \geq b$ if and only if $a - m \geq |x - m|$.

   b $a \geq x \geq b$ if and only if $a - b \geq |2x - (a + b)|$

   c $a > x > b$ if and only if $a - m > |x - m|$.

   d $a > x > b$ if and only if $a - b > |2x - (a + b)|$

CHAPTER 5

Max and Min Operations

5.1 Max Operation

1. In this section, we will define the max operation and its notation. We will also provide some examples of the max operation. Additionally, we will prove some basic properties of the max operation. Finally, we will use associativity to generalize the max operation, allowing for more than two arguments.

2. **Definition of the max operation.** The maximum of two real numbers x, y is x if $x \geq y$ and y if $x < y$. We denote the maximum of two real numbers x, y by $\max(x, y)$

3. Let us find the maximum of the following pairs of real numbers.

4. 2, 3

    Solution Since $2 < 3$ then $\max(2, 3) = 3$.

5. 5, 7

    Solution Since $5 < 7$ then $\max(5, 7) = 7$.

6. 1, 1

    Solution Since $1 \geq 1$ then $\max(1, 1) = 1$.

7. **Properties of the Max Operation.** Here, we prove the following properties of the max operation.

    - $\max(x, x) = x$
    - $\max(x, y) = \max(y, x)$

- $\max(x, \max(y, z)) = \max(\max(x, y), z)$
- If $z \geq 0$ then $z \max(x, y) = \max(xz, yz)$.

8   An alternative definition of the max operation. For all real numbers x, y

1. $\max(x, y) = x$ if $x > y$
2. $\max(x, y) = y$ if $x \leq y$

Proof   When $x > y$, we have that $x \geq y$. Hence $\max(x, y) = x$.

When $x = y$, we have that $x \geq y$. Hence $\max(x, y) = x = y$.

When $x < y$, $\max(x, y) = y$. We can combine the cases $x = y$ and $x < y$: when $x \leq y$, $\max(x, y) = y$. ∎

9   $\max(x, x) = x$

Proof   Since $x \geq x$ the max operation gives $\max(x, x) = x$. ∎

10   $\max(x, y) = \max(y, x)$

(Commutativity of the Max Operation)

Proof   When $x \geq y$, $\max(x, y) = x$ and using the alternative definition we get $\max(y, x) = x$. Thus $\max(x, y) = \max(y, x)$. When $x < y$, $\max(x, y) = y$ and using the alternative definition we get $\max(y, x) = y$. Thus $\max(x, y) = \max(y, x)$. Therefore in both cases $\max(x, y) = \max(y, x)$. ∎

11   If $x \geq y$ and $x \geq z$ then $\max(x, \max(y, z)) = x$.

Proof   There are two cases: when $y \geq z$ and when $y < z$.

When $y \geq z$, $\max(y, z) = y$. Hence $\max(x, \max(y, z)) = \max(x, y)$. Thus $\max(x, \max(y, z)) = x$.

When $y < z$, $\max(y, z) = z$. Hence $\max(x, \max(y, z)) = \max(x, z)$. Thus $\max(x, \max(y, z)) = x$. ∎

12   $\max(x, \max(y, z)) = \max(\max(x, y), z)$

(Associativity of the Max Operation)

Proof Let us consider the case where x is the greatest. Since x is the greatest it follows that $x \geq y$ and $x \geq z$. Thus,

$$\begin{aligned} \max(\max(x, y), z) &= \max(x, z) \\ &= x \end{aligned} \quad (1)$$

From the previous result we obtain

$$\max(x, \max(y, z)) = x \quad (2)$$

Hence from (1) and (2) we get

$$\max(x, \max(y, z)) = \max(\max(x, y), z)$$

Let us consider the case where y is the greatest. Since y is the greatest it follows that $y \geq x$ and $y \geq z$. Thus

$$\begin{aligned} \max(\max(x, y), z) &= \max(y, z) \\ &= y \end{aligned} \quad (1)$$

And,

$$\begin{aligned} \max(x, \max(y, z)) &= \max(x, y) \\ &= y \end{aligned} \quad (2)$$

Hence from (1) and (2) we get

$$\max(x, \max(y, z)) = \max(\max(x, y), z)$$

Let us consider the case where z is the greatest. Since z is the greatest it follows that $z \geq x$ and $z \geq y$, then

$$\max(x, z) = z$$
$$\max(y, z) = z$$

Thus

$$\begin{aligned} \max(x, \max(y, z)) &= \max(x, z) \\ &= z \end{aligned} \quad (1)$$

From the previous result we obtain

$$\max(z, \max(x, y)) = z$$

Due to commutativity we have

$$\max(\max(x,y),z) = z \qquad (2)$$

Hence from (1) and (2) we get

$$\max(x, \max(y,z)) = \max(\max(x,y), z)$$

In all three cases we get

$$\max(x, \max(y,z)) = \max(\max(x,y), z)$$

∎

13 If $z \geq 0$ then $z \max(x,y) = \max(xz, yz)$

(Distribution of Multiplication over the Max operation)

Proof There are two cases: when $x \geq y$ and when $x < y$.

When $x \geq y$, $\max(x,y) = x$. Multiplying the equation by z, we obtain

$$z \max(x,y) = zx.$$

Multiplying the inequality by z, we obtain $zx \geq zy$. Hence $\max(zx, zy) = zx$. Thus $z \max(x,y) = \max(zx, zy)$.

When $x < y$, $\max(x,y) = y$. Multiplying the equation by z, we obtain

$$z \max(x,y) = zy.$$

Multiplying the inequality by z, we obtain $zx < zy$. Hence $\max(zx, zy) = zy$. Thus $z \max(x,y) = \max(zx, zy)$. In both cases $z \max(x,y) = \max(xz, yz)$. ∎

14 **Max operation for more than two arguments.** Since the max operation is associative,

$$\max(x, \max(y,z)) = \max(\max(x,y), z)$$

Hence we may simply write $\max(x, y, z)$ without ambiguity. In the case of four arguments we have

$$\max(w, \max(x, y, z)) = \max(\max(w, x, y), z)$$

In general, for any finite list of real numbers, $x_1, x_2, \ldots, x_n$ we may write

$$\max(x_1, x_2, \ldots, x_n)$$

to denote the greatest member of the list.

Exercises 5.1

Let x be a real number and let n be an integer

1 Compute the following

    a $\max(1, 2)$

    b $\max(-2, -2)$

    c $\max(-3, -4)$

    d $\max(x^2, -x^2)$

2 Compute the following

    a $\max(1, 2, 3)$

    b $\max(-3, -4, -3)$

    c $\max(-2, -2, -5, 0, 3)$

    d $\max(x + 1, x + 2, x - 3)$

3 Prove the following

    a If $n > 3$ then $\max(n!, n^2) = n!$

    b If $n > 5$ then $\max(n!, n^3) = n!$

5.2 Min Operation

1 In this section, we will define the min operation and its notation. We will also provide some examples of the min operation. Additionally, we will prove some basic properties of the min operation. Finally, we will use associativity to generalize the min operation, allowing for more than two arguments.

2 **Definition of the min operation.** The minimum of two real numbers x, y is x if $x \leq y$ and y if $x > y$. We denote the minimum of two real numbers x, y by $\min(x, y)$.

3 Let us find the minimum of the following pairs of real numbers.

4 2, 3

    Solution Since $2 \leq 3$, $\min(2, 3) = 2$.

Chapter 5 Max and Min Operations

5 1, 1

Solution Since $1 \leq 1$, $\min(1, 1) = 1$.

6 $-5, -7$

Solution Since $-5 > -7$, $\min(-5, -7) = -7$.

7 $5.5, 3.6$

Solution Since $5.5 > 3.6$, $\min(5.5, 3.6) = 3.6$.

8 **Properties of the Min Operation.** Here, we prove the following properties of the min operation.

- $\min(x, x) = x$
- $-\min(x, y) = \max(-x, -y)$
- $\min(x, y) = \min(y, x)$
- $\min(x, \min(y, z)) = \min(\min(x, y), z)$
- If $z \geq 0$ then $z \min(x, y) = \min(xz, yz)$.

9 Let x, y, z be real numbers.

10 $\min(x, x) = x$

Proof Since $x \leq x$ the min operation gives $\min(x, x) = x$. ∎

11 $-\min(x, y) = \max(-x, -y)$

Proof There are two cases: when $x \leq y$ and when $x > y$.

When $x \leq y$, $\min(x, y) = x$. Multiplying both sides of the equation by -1 we obtain $-\min(x, y) = -x$. Multiplying both sides of the inequality by -1 we obtain $-x \geq -y$. Hence $\max(-x, -y) = -x$. Therefore $-\min(x, y) = \max(-x, -y)$.

When $x > y$, $\min(x, y) = y$. Multiplying both sides of the equation by -1 we obtain $-\min(x, y) = -y$. Multiplying both sides of the inequality by -1 we obtain $-x < -y$. Hence $\max(-x, -y) = -y$. Therefore $-\min(x, y) = \max(-x, -y)$. ∎

12   $\min(x, y) = \min(y, x)$

(Commutativity of the Min Operation)

Proof Due to commutativity of the max operation we have
$$\max(-x, -y) = \max(-y, -x)$$
$$-\min(x, y) = -\min(y, x)$$
Divide both sides by -1
$$\min(x, y) = \min(y, x)$$

∎

13   $\min(x, \min(y, z)) = \min(\min(x, y), z)$

(Associativity of the Min Operation)

Proof Due to associativity of the max operation,
$$\max(-x, \max(-y, -z)) = \max(\max(-x, -y), -z)$$
$$\max(-x, -\min(y, z)) = \max(-\min(x, y), -z)$$
$$-\min(x, \min(y, z)) = -\min(\min(x, y), z)$$
Divide both sides by -1
$$\min(x, \min(y, z)) = \min(\min(x, y), z)$$

∎

14   If $z \geq 0$ then $z \min(x, y) = \min(xz, yz)$

(Distribution of Multiplication over the Min operation)

Proof Due to Distribution of Multiplication over Max Operation,
$$z \max(-x, -y) = \max(-xz, -yz)$$
$$-z \min(x, y) = -\min(xz, yz)$$
Divide both sides by -1
$$z \min(x, y) = \min(xz, yz)$$

∎

15   **Min operation for more than two arguments.** Since the min operation is associative, for any finite list of real numbers, $x_1, x_2, \ldots, x_n$ we may write
$$\min(x_1, x_2, \ldots, x_n)$$
to denote the least member of the list.

Exercises 5.2

Let x be a real number and let n be an integer

1 Compute the following

    a $\min(1, 2)$

    b $\min(-2, -2)$

    c $\min(-3, -4)$

    d $\min(x^2, -x^2)$

2 Compute the following

    a $\min(1, 2, 3)$

    b $\min(-3, -4, -3)$

    c $\min(-2, -2, -5, 0, 3)$

    d $\min(x + 1, x + 2, x - 3)$

3 Prove the following

    a If $n > 3$ then $\min(n!, n^2) = n^2$

    b If $n > 5$ then $\min(n!, n^3) = n^3$

5.3 Max and Min Operations and Absolute Value

1 In this section, we write formulas for the max and min operations using absolute value:

- $\max(x, y) = \dfrac{x + y + |x - y|}{2}$

- $\min(x, y) = \dfrac{x + y - |x - y|}{2}$

2 Let x, y, z be real numbers.

3 $\max(x, y) + \min(x, y) = x + y$

5.3 Max and Min Operations and Absolute Value

Proof There are two cases: when $x \geq y$ and when $x < y$.

When $x \geq y$, $\max(x, y) = x$ and $\min(x, y) = y$. Adding both equations, we obtain $\max(x, y) + \min(x, y) = x + y$.

When $x < y$, $\max(x, y) = y$ and $\min(x, y) = x$. Adding both equations, we obtain $\max(x, y) + \min(x, y) = y + x$. Hence in both cases $\max(x, y) + \min(x, y) = x + y$ ∎

4 If $x \geq y$ then $\max(2x, 2y) - (x + y) = |x - y|$

Proof

$$x \geq y \qquad (1)$$

Multiply both sides by 2

$$2x \geq 2y$$

The definition of max operation gives

$$\max(2x, 2y) = 2x$$

Subtract $x + y$ from both sides

$$\max(2x, 2y) - (x + y) = 2x - (x + y)$$
$$= x - y \qquad (2)$$

From (1) we have

$$x - y \geq 0$$

By the definition of absolute value

$$|x - y| = x - y$$

Using the above equation we can substitute for $(x - y)$ in (2)

$$\max(2x, 2y) - (x + y) = |x - y|$$

∎

5 If $x < y$ then $\max(2x, 2y) - (x + y) = |x - y|$

Proof

$$x < y \tag{1}$$

Multiply both sides by 2

$$2x < 2y$$

The definition of max operation gives

$$\max(2x, 2y) = 2y$$

Subtract $x + y$ from both sides

$$\max(2x, 2y) - (x + y) = 2y - (x + y)$$
$$= y - x \tag{2}$$

From (1) we have

$$x - y < 0$$

By the definition of absolute value

$$|x - y| = -(x - y)$$
$$= y - x$$

Using the above equation we can substitute for $(x - y)$ in (2)

$$\max(2x, 2y) - (x + y) = |x - y|$$

∎

6   $\max(2x, 2y) - (x + y) = |x - y|$

Proof It is either $x \geq y$ or $x < y$. In each case $\max(2x, 2y) - (x+y) = |x-y|$. ∎

7
$$\max(x, y) = \frac{x + y + |x - y|}{2}$$

(Formula for Max Operation)

5.3 Max and Min Operations and Absolute Value

Proof From the previous result,

$$\max(2x, 2y) - (x+y) = |x-y|$$
$$2\max(x,y) - (x+y) = |x-y|$$
$$2\max(x,y) = x+y+|x-y|$$

Divide both sides by 2

$$\max(x,y) = \frac{x+y+|x-y|}{2}$$

∎

8
$$\min(x,y) = \frac{x+y-|x-y|}{2}$$

(Formula for Min Operation)

Proof From the formula for max operation

$$\max(-x, -y) = \frac{(-x)+(-y)+|(-x)-(-y)|}{2}$$
$$-\min(x,y) = \frac{-x-y+|-x+y|}{2}$$
$$= \frac{-(x+y)+|y-x|}{2}$$
$$= \frac{-(x+y)+|x-y|}{2}$$
$$= -\frac{(x+y)-|x-y|}{2}$$

Divide both sides by -1

$$\min(x,y) = \frac{x+y-|x-y|}{2}$$

∎

CHAPTER 6
Floor and Ceiling

6.1 Floor

1. In this section, we define the floor of a real number and its notation. We also provide some examples of the floor function.

2. **Definition of floor.** The floor of a real number x, is the greatest integer less than or equal to x. It is a real number n satisfying the following

   1. n is an integer and $n \leq x$
   2. If m is an integer and $m \leq x$ then $m \leq n$.

3. While writing a proof, we prove that n is the floor of x by showing that n satisfies both conditions above.

4. **Notation of floor.** The floor of a real number x is denoted as $\lfloor x \rfloor$.

5. **Examples of floor.** Let us find the floor of some real numbers.

6. $\lfloor 10 \rfloor$.

   Solution The integers not greater than 10 are
   $$10, 9, 8, 7, 6, \ldots$$
   Since 10 is the greatest of these integers it follows that
   $$\lfloor 10 \rfloor = 10$$

7. For all integers n, the floor of n is the same as n.

8. Other examples are shown below.

   - $\lfloor 0 \rfloor = 0$

- $\lfloor 1 \rfloor = 1$
- $\lfloor 2 \rfloor = 2$
- $\lfloor -1 \rfloor = -1$
- $\lfloor -2 \rfloor = -2$

9 $\lfloor 5.5 \rfloor$.

Solution The integers not greater than 5.5 are

$$5, 4, 3, 2, \ldots$$

The integer 5 is the greatest of these. Hence

$$\lfloor 5.5 \rfloor = 5$$

10 $\lfloor -3.2 \rfloor$.

Solution The integers not greater than -3.2 are

$$-4, -5, -6, \ldots$$

The integer -4 is the greatest of these. Hence

$$\lfloor -3.2 \rfloor = -4.$$

Exercises 6.1

Compute the following

1 $\lfloor 2 \rfloor$

2 $\lfloor 4.4 \rfloor$

3 $\lfloor -2.3 \rfloor$

4 $\lfloor -5.9 \rfloor$

6.2 Basic Properties of the Floor Function

1. In this section, we prove the following properties of the floor function.

    - $\lfloor x \rfloor = x$ if and only if x is an integer.
    - $n \le x$ if and only if $n \le \lfloor x \rfloor$
    - $\lfloor x + n \rfloor = \lfloor x \rfloor + n$
    - $x - 1 < \lfloor x \rfloor \le x$
    - $\lfloor x \rfloor \le x < \lfloor x \rfloor + 1$
    - If $x \ge y$ then $\lfloor x \rfloor \ge \lfloor y \rfloor$ (Floor of Inequality)
    - If $n > 0$ then
    $$\left\lfloor \frac{x}{n} \right\rfloor = \left\lfloor \frac{\lfloor x \rfloor}{n} \right\rfloor$$

2. Let x, y be real numbers. Let n be an integer.

3. $\lfloor x \rfloor = x$ if and only if x is an integer.

    Proof Let $\lfloor x \rfloor = x$. Since by definition $\lfloor x \rfloor$ is an integer, it follows that x is an integer.

    Conversely, let x be an integer. Thus $x \le x$. Let m be an integer such that $m \ge x$. Thus $m \ge x$. It follows that $x = \lfloor x \rfloor$. ∎

4. $n \le x$ if and only if $n \le \lfloor x \rfloor$

    Proof Given $n \le x$, since n is an integer, the definition of $\lfloor x \rfloor$ gives that $n \le \lfloor x \rfloor$. Conversely, let $n \le \lfloor x \rfloor$. We know that $\lfloor x \rfloor \le x$. If we apply transitivity we get $n \le x$. ∎

5. $n > \lfloor x \rfloor$ if and only if $n > x$

    Proof This is the contrapositive of the previous result ∎

6. $\lfloor x \rfloor > x - 1$

Proof

We know that

$$1 > 0$$

add $\lfloor x \rfloor$ to both sides

$$\lfloor x \rfloor + 1 > \lfloor x \rfloor$$

Since the LHS is an integer, we can apply the previous result

$$\lfloor x \rfloor + 1 > x$$
$$\lfloor x \rfloor > x - 1$$

∎

7 $\lfloor x + n \rfloor = \lfloor x \rfloor + n$

Proof

By definition of the floor function,

$$\lfloor x + n \rfloor \leq x + n$$
$$\lfloor x + n \rfloor - n \leq x$$

Since the LHS is an integer, from a previous lemma we can write

$$\lfloor x + n \rfloor - n \leq \lfloor x \rfloor$$
$$\lfloor x + n \rfloor \leq \lfloor x \rfloor + n \qquad (1)$$

By definition of the floor function

$$\lfloor x \rfloor \leq x$$

Add n to both sides

$$\lfloor x \rfloor + n \leq x + n$$

Since the LHS is an integer, from a previous lemma we can write

$$\lfloor x \rfloor + n \leq \lfloor x + n \rfloor \qquad (2)$$

From (1) and (2), we get

$$\lfloor x + n \rfloor = \lfloor x \rfloor + n$$

∎

8 $\lfloor x - n \rfloor = \lfloor x \rfloor - n$

Proof From the previous result we have that for any integer m

$$\lfloor x + m \rfloor = \lfloor x \rfloor + m$$

Put $m = -n$

$$\lfloor x - n \rfloor = \lfloor x \rfloor - n$$

■

9 $x - 1 < \lfloor x \rfloor \leq x$

Proof

By definition of the floor function

$$\lfloor x \rfloor \leq x$$

From a previous result we know that

$$x - 1 < \lfloor x \rfloor$$

Chain the above inequalities

$$x - 1 < \lfloor x \rfloor \leq x$$

■

10 If $x - 1 < n \leq x$ then $n = \lfloor x \rfloor$.

Proof

$$x - 1 < n \leq x$$

Split the chained inequality.

$$x - 1 < n \tag{1}$$
$$n \leq x \tag{2}$$

From (2) and the definition of the floor function, we get

$$n \le \lfloor x \rfloor \tag{3}$$

From (1), we get

$$n > x - 1$$

From a previous result, we get

$$n > \lfloor x - 1 \rfloor$$
$$n > \lfloor x \rfloor - 1$$
$$n \ge \lfloor x \rfloor \tag{4}$$

From (3) and (4), we get

$$n = \lfloor x \rfloor$$

∎

11 $\lfloor x \rfloor \le x < \lfloor x \rfloor + 1$

Proof

By definition of the floor function

$$\lfloor x \rfloor \le x \tag{1}$$

From a previous result we know that

$$x - 1 < \lfloor x \rfloor$$

Add 1 to both sides

$$x < \lfloor x \rfloor + 1 \tag{2}$$

Chain (1) and (2)

$$\lfloor x \rfloor \le x < \lfloor x \rfloor + 1$$

∎

12   If $n \le x < n + 1$ then $n = \lfloor x \rfloor$.

6.2 Basic Properties of the Floor Function

Proof

$$n \leq x < n + 1$$

Split the chained inequality.

$$n \leq x \quad (1)$$
$$x < n + 1 \quad (2)$$

From (1) and the definition of the floor function, we get

$$n \leq \lfloor x \rfloor \quad (3)$$

From (2), we get

$$n > x - 1$$

From a previous result, we get

$$n > \lfloor x - 1 \rfloor$$
$$n > \lfloor x \rfloor - 1$$
$$n \geq \lfloor x \rfloor \quad (4)$$

From (3) and (4), we get

$$n = \lfloor x \rfloor$$

∎

13 If $x \geq y$ then $\lfloor x \rfloor \geq \lfloor y \rfloor$

(Floor of Inequality)

Proof It is either $y < \lfloor x \rfloor$ or $y \geq \lfloor x \rfloor$. In the first case, $\lfloor x \rfloor > y$. Apply transitivity to $\lfloor x \rfloor > y$ and $y \geq \lfloor y \rfloor$. We get $\lfloor x \rfloor > \lfloor y \rfloor$. Thus $\lfloor x \rfloor \geq \lfloor y \rfloor$.

In the second case, $\lfloor x \rfloor \leq y$. We prove this case by contradiction. Assume $\lfloor x \rfloor < \lfloor y \rfloor$. Thus $\lfloor y \rfloor \geq \lfloor x \rfloor + 1$. Since $\lfloor x \rfloor + 1 > x$, we may write $\lfloor y \rfloor > x$. Since $y \geq \lfloor y \rfloor$, we have $y > x$. Thus $x < y$ and $x \geq y$. This contradiction means the assumption $\lfloor x \rfloor < \lfloor y \rfloor$ is wrong. Thus $\lfloor x \rfloor \geq \lfloor y \rfloor$. ∎

14 If $\lfloor x \rfloor < \lfloor y \rfloor$ then $x < y$

Proof This is the contrapositive of the previous result. ∎

Chapter 6 Floor and Ceiling

15 If $n > 0$ then
$$\left\lfloor \frac{x}{n} \right\rfloor + 1 > \frac{\lfloor x \rfloor}{n}$$

Proof From a previous result
$$\left\lfloor \frac{x}{n} \right\rfloor > \frac{x}{n} - 1$$

Multiply both sides by n
$$n \left\lfloor \frac{x}{n} \right\rfloor > x - n$$
$$n \left\lfloor \frac{x}{n} \right\rfloor + n > x$$

Since $x \geq \lfloor x \rfloor$, it follows that
$$n \left\lfloor \frac{x}{n} \right\rfloor + n > \lfloor x \rfloor$$

Divide both sides by n
$$\left\lfloor \frac{x}{n} \right\rfloor + 1 > \frac{\lfloor x \rfloor}{n}$$

∎

16 If $n > 0$ then
$$\frac{\lfloor x \rfloor}{n} \geq \left\lfloor \frac{x}{n} \right\rfloor$$

Proof By definition of the floor function,
$$\frac{x}{n} \geq \left\lfloor \frac{x}{n} \right\rfloor$$

Multiply both sides by n
$$x \geq n \cdot \left\lfloor \frac{x}{n} \right\rfloor$$

Since the RHS is an integer, we may write
$$\lfloor x \rfloor \geq n \cdot \left\lfloor \frac{x}{n} \right\rfloor$$

Divide both sides by n
$$\frac{\lfloor x \rfloor}{n} \geq \left\lfloor \frac{x}{n} \right\rfloor$$

∎

17 If $n > 0$ then
$$\left\lfloor \frac{x}{n} \right\rfloor = \left\lfloor \frac{\lfloor x \rfloor}{n} \right\rfloor$$

Proof From previous results, $\left\lfloor \frac{x}{n} \right\rfloor + 1 > \frac{\lfloor x \rfloor}{n}$ and $\frac{\lfloor x \rfloor}{n} \geq \left\lfloor \frac{x}{n} \right\rfloor$. Thus,
$$\left\lfloor \frac{x}{n} \right\rfloor + 1 > \frac{\lfloor x \rfloor}{n} \geq \left\lfloor \frac{x}{n} \right\rfloor$$

It follows that
$$\left\lfloor \frac{x}{n} \right\rfloor = \left\lfloor \frac{\lfloor x \rfloor}{n} \right\rfloor$$

■

18 If $n > 0$ then
$$\left\lfloor \frac{x+m}{n} \right\rfloor = \left\lfloor \frac{\lfloor x \rfloor + m}{n} \right\rfloor$$

Proof Let y be a real number. By a previous result we have:
$$\left\lfloor \frac{y}{n} \right\rfloor = \left\lfloor \frac{\lfloor y \rfloor}{n} \right\rfloor$$

Put $y = x + m$
$$\left\lfloor \frac{x+m}{n} \right\rfloor = \left\lfloor \frac{\lfloor x+m \rfloor}{n} \right\rfloor$$

Since m is an integer, we may write:
$$\left\lfloor \frac{x+m}{n} \right\rfloor = \left\lfloor \frac{\lfloor x \rfloor + m}{n} \right\rfloor$$

■

Exercises 6.2

1 Let x be a real number. Prove the following.

  a $x \geq \lfloor x \rfloor > x - 1$

  b $\lfloor x \rfloor + 1 > x \geq \lfloor x \rfloor$

  c $\lfloor x \rfloor + 1 > x \geq \lfloor x \rfloor > x - 1$

  d $x - 1 < \lfloor x \rfloor \leq x < \lfloor x \rfloor + 1$

e If $x \leq y$ then $\lfloor x \rfloor \leq \lfloor y \rfloor$
(Floor of Inequality)

f $x \geq n$ if and only if $\lfloor x \rfloor \geq n$

g $\lfloor x \rfloor < n$ if and only if $x < n$

2 Let a, b, x be real numbers. Prove the following.

a If $a \leq x \leq b$ then $\lfloor a \rfloor \leq \lfloor x \rfloor \leq \lfloor b \rfloor$.

b If $a \geq x \geq b$ then $\lfloor a \rfloor \geq \lfloor x \rfloor \geq \lfloor b \rfloor$.

3 Let a, b, x be real numbers. Prove the following.

a If $a < x < b$ then $\lfloor a \rfloor < \lfloor x \rfloor < \lfloor b \rfloor$.

b If $a > x > b$ then $\lfloor a \rfloor > \lfloor x \rfloor > \lfloor b \rfloor$.

4 Let x be a real number. Prove the following.

a $\lfloor \lfloor x \rfloor \rfloor = \lfloor x \rfloor$

b $\lfloor x \rfloor + \lfloor y \rfloor \leq \lfloor x + y \rfloor \leq \lfloor x \rfloor + \lfloor y \rfloor + 1$

5 Let x be a real number. Let n and $a_1, \ldots, a_n$ be positive integers. Prove the following.

a $\left\lfloor \dfrac{1}{a_1} \cdot x \right\rfloor = \left\lfloor \dfrac{x}{a_1} \right\rfloor$

b $\left\lfloor \dfrac{1}{a_1} \cdot \left\lfloor \dfrac{1}{a_2} \cdot x \right\rfloor \right\rfloor = \left\lfloor \dfrac{x}{a_1 a_2} \right\rfloor$

c $\left\lfloor \dfrac{1}{a_1} \cdot \left\lfloor \dfrac{1}{a_2} \cdot \left\lfloor \dfrac{1}{a_3} \cdot x \right\rfloor \right\rfloor \right\rfloor = \left\lfloor \dfrac{x}{a_1 a_2 a_3} \right\rfloor$

d $\left\lfloor \dfrac{1}{a_1} \cdot \left\lfloor \dfrac{1}{a_2} \cdot \left\lfloor \dfrac{1}{a_3} \cdot \left\lfloor \dfrac{1}{a_4} \cdot x \right\rfloor \right\rfloor \right\rfloor \right\rfloor = \left\lfloor \dfrac{x}{a_1 a_2 a_3 a_4} \right\rfloor$

6.3 Ceiling

1. In this section, we define the ceiling of a real number and its notation. We also provide some examples of the ceiling function.

2. **Definition of ceiling.** The ceiling of a real number x, is the least integer greater than or equal to x. It is a real number n satisfying the following

    1. n is an integer and $n \geq x$
    2. If m is an integer and $m \geq x$ then $m \geq n$.

3. While writing a proof, we prove that n is the ceiling of x by showing that n satisfies both conditions above.

4. **Notation of ceiling.** The ceiling of a real number x is denoted as $\lceil x \rceil$.

5. **Examples of ceiling.** Let us compute the ceiling of some real numbers as examples.

6. $\lceil 10 \rceil$

    Solution The integers not less than 10 are

    $$10, 11, 12, 13, 14, \ldots$$

    Since 10 is the least of these integers it follows that

    $$\lceil 10 \rceil = 10$$

7. For all integers n, the ceiling of n is the same as n.

8. Other examples are shown below.
    - $\lceil 0 \rceil = 0$
    - $\lceil 1 \rceil = 1$
    - $\lceil 2 \rceil = 2$
    - $\lceil -1 \rceil = -1$
    - $\lceil -2 \rceil = -2$

9. $\lceil 5.5 \rceil$.

Solution The integers not less than 5.5 are

$$6, 7, 8, 9, \ldots$$

The integer 6 is the least of these. Hence

$$\lceil 5.5 \rceil = 6$$

Exercises 6.3

Compute the following.

1 $\lceil 2 \rceil$

2 $\lceil 4.8 \rceil$

3 $\lceil -2.6 \rceil$

4 $\lceil -0.5 \rceil$

6.4 Basic Properties of the Ceiling function

1 In this section, we establish a formula that converts floor to ceiling: for all real numbers x,

$$\lfloor x \rfloor = -\lceil -x \rceil.$$

Using this formula, we rephrase theorems we have previously proven for floor so that they are also proven for ceiling. Some of the theorems we prove are listed below.

- $\lceil x \rceil = x$ if and only if x is an integer.
- $n \geq x$ if and only if $n \geq \lceil x \rceil$
- $\lceil x + n \rceil = \lceil x \rceil + n$
- $x + 1 > \lceil x \rceil \geq x$
- $\lceil x \rceil \geq x > \lceil x \rceil - 1$
- If $x \leq y$ then $\lceil x \rceil \leq \lceil y \rceil$ (Ceiling of Inequality)
- If $n > 0$ then

$$\left\lceil \frac{x}{n} \right\rceil = \left\lceil \frac{\lceil x \rceil}{n} \right\rceil$$

2 Let x, y be real numbers. Let n be an integer.

3 $\lfloor x \rfloor = -\lceil -x \rceil$.

Proof We know that $\lfloor x \rfloor \leq x$. By negating the inequality, we get $-\lfloor x \rfloor \geq -x$. Let m be an integer such that $m \geq -x$. Thus $-m \leq x$. Hence $-m \leq \lfloor x \rfloor$. It follows that $m \geq -\lfloor x \rfloor$. Thus $-\lfloor x \rfloor = \lceil -x \rceil$. Hence $\lfloor x \rfloor = -\lceil -x \rceil$. ∎

4 $\lceil x \rceil = -\lfloor -x \rfloor$.

Proof From the previous result we know that for any real number y

$$-\lfloor y \rfloor = \lceil -y \rceil$$

Put $y = -x$

$$-\lfloor -x \rfloor = \lceil -(-x) \rceil$$
$$-\lfloor -x \rfloor = \lceil x \rceil$$

∎

5 $\lceil x \rceil = x$ if and only if x is an integer.

Proof We know that $\lfloor y \rfloor = y$ if and only if y is an integer. Hence $-\lceil -y \rceil = y$ iff y is an integer. Let $y = -x$. Thus $-\lceil x \rceil = -x$ iff $-x$ is an integer. By cancelling -1 we have $\lceil x \rceil = x$ iff $-x$ is an integer. We know that x is an integer iff $-x$ is an integer. Therefore $\lceil x \rceil = x$ iff x is an integer. ∎

6 $n \geq x$ if and only if $n \geq \lceil x \rceil$

Proof Given $n \geq x$, since n is an integer, the definition of $\lceil x \rceil$ gives that $n \geq \lceil x \rceil$. Conversely, let $n \geq \lceil x \rceil$. We know that $\lceil x \rceil \geq x$. If we apply transitivity we get $n \geq x$. ∎

7 $n < \lceil x \rceil$ if and only if $n < x$

Proof This is the contrapositive of the previous result ∎

8 $\lceil x \rceil < x + 1$

Proof

We know that

$$-1 < 0$$

add $\lceil x \rceil$ to both sides

$$\lceil x \rceil - 1 < \lceil x \rceil$$

Since the LHS is an integer, we can apply the previous result

$$\lceil x \rceil - 1 < x$$
$$\lceil x \rceil < x + 1$$

∎

9 $\lceil x + n \rceil = \lceil x \rceil + n$

Proof We know that

$$\lfloor -x - n \rfloor = \lfloor -x \rfloor - n$$

Thus,

$$\lfloor -(x + n) \rfloor = -\lceil x \rceil - n$$
$$-\lceil x + n \rceil = -\lceil x \rceil - n$$
$$\lceil x + n \rceil = \lceil x \rceil + n$$

∎

10 $\lceil x - n \rceil = \lceil x \rceil - n$

Proof From the previous result we have that for any integer m

$$\lceil x + m \rceil = \lceil x \rceil + m$$

Put $m = -n$

$$\lceil x - n \rceil = \lceil x \rceil - n$$

∎

11   $x + 1 > \lceil x \rceil \geq x$

Proof

By definition of the ceiling function

$$\lceil x \rceil \geq x$$

From a previous result we know that

$$x + 1 > \lceil x \rceil$$

Chain the above inequalities

$$x + 1 > \lceil x \rceil \geq x$$

∎

12   If $x + 1 > n \geq x$ then $n = \lceil x \rceil$.

Proof

$$x + 1 > n \geq x$$

Split the chained inequality.

$$x + 1 > n \quad (1)$$
$$n \geq x \quad (2)$$

From (2) and the definition of the ceiling function, we get

$$n \geq \lceil x \rceil \quad (3)$$

From (1), we get

$$n < x + 1$$

From a previous result, we get

$$n < \lceil x + 1 \rceil$$
$$n < \lceil x \rceil + 1$$
$$n \leq \lceil x \rceil \quad (4)$$

From (3) and (4), we get

$$n = \lceil x \rceil$$

∎

13   $\lceil x \rceil \geq x > \lceil x \rceil - 1$

Proof

By definition of the ceiling function

$$\lceil x \rceil \geq x \qquad (1)$$

From a previous result we know that

$$x + 1 > \lceil x \rceil$$

Subtract 1 from both sides

$$x > \lceil x \rceil - 1 \qquad (2)$$

Chain (1) and (2)

$$\lceil x \rceil \geq x > \lceil x \rceil - 1$$

∎

14   If $n \geq x > n - 1$ then $n = \lceil x \rceil$.

Proof

$$n \geq x > n - 1$$

Split the chained inequality.

$$n \geq x \qquad (1)$$
$$x > n - 1 \qquad (2)$$

From (1) and the definition of the floor function, we get

$$n \geq \lceil x \rceil \qquad (3)$$

From (2), we get

$$n < x + 1$$

From a previous result, we get

$$n < \lceil x + 1 \rceil$$
$$n < \lceil x \rceil + 1$$
$$n \leq \lceil x \rceil \qquad (4)$$

From (3) and (4), we get

$$n = \lceil x \rceil$$

∎

6.4 Basic Properties of the Ceiling function

15. If $x \leq y$ then $\lceil x \rceil \leq \lceil y \rceil$

    (Ceiling of Inequality)

    Proof Since $x \leq y$, it follows that $-x \geq -y$. Thus $\lfloor -x \rfloor \geq \lfloor -y \rfloor$. Hence $-\lceil x \rceil \geq -\lceil y \rceil$. Thus $\lceil x \rceil \leq \lceil y \rceil$. ∎

16. If $\lceil x \rceil > \lceil y \rceil$ then $x > y$

    Proof This is the contrapositive of the previous result. ∎

17. If $n > 0$ then
    $$\left\lceil \frac{x}{n} \right\rceil = \left\lceil \frac{\lceil x \rceil}{n} \right\rceil$$

    Proof From a previous result,
    $$\left\lfloor \frac{-x}{n} \right\rfloor = \left\lfloor \frac{\lfloor -x \rfloor}{n} \right\rfloor$$
    $$\left\lfloor -\frac{x}{n} \right\rfloor = \left\lfloor \frac{-\lceil x \rceil}{n} \right\rfloor$$
    $$-\left\lceil \frac{x}{n} \right\rceil = \left\lfloor -\frac{\lceil x \rceil}{n} \right\rfloor$$
    $$-\left\lceil \frac{x}{n} \right\rceil = -\left\lceil \frac{\lceil x \rceil}{n} \right\rceil$$
    $$\left\lceil \frac{x}{n} \right\rceil = \left\lceil \frac{\lceil x \rceil}{n} \right\rceil$$
    ∎

18. If $n > 0$ then
    $$\left\lceil \frac{x+m}{n} \right\rceil = \left\lceil \frac{\lceil x \rceil + m}{n} \right\rceil$$

    Proof Let y be a real number. By a previous result we have:
    $$\left\lceil \frac{y}{n} \right\rceil = \left\lceil \frac{\lceil y \rceil}{n} \right\rceil$$

    Put $y = x + m$
    $$\left\lceil \frac{x+m}{n} \right\rceil = \left\lceil \frac{\lceil x+m \rceil}{n} \right\rceil$$

    Since m is an integer, we may write:
    $$\left\lceil \frac{x+m}{n} \right\rceil = \left\lceil \frac{\lceil x \rceil + m}{n} \right\rceil$$
    ∎

Exercises 6.4

1 Let x be a real number. Prove the following.

    a $x \leq \lceil x \rceil < x + 1$

    b $\lceil x \rceil - 1 < x \leq \lceil x \rceil$

    c $\lceil x \rceil - 1 < x \leq \lceil x \rceil < x + 1$

    d $x + 1 > \lceil x \rceil \geq x > \lceil x \rceil - 1$

    e If $x \geq y$ then $\lceil x \rceil \geq \lceil y \rceil$
        (Ceiling of Inequality)

    f $x \leq n$ if and only if $\lceil x \rceil \leq n$

    g $\lceil x \rceil > n$ if and only if $x > n$

2 Let a, b, x be real numbers. Prove the following.

    a If $a \leq x \leq b$ then $\lceil a \rceil \leq \lceil x \rceil \leq \lceil b \rceil$.

    b If $a \geq x \geq b$ then $\lceil a \rceil \geq \lceil x \rceil \geq \lceil b \rceil$.

3 Let a, b, x be real numbers. Prove the following.

    a If $a < x < b$ then $\lceil a \rceil < \lceil x \rceil < \lceil b \rceil$.

    b If $a > x > b$ then $\lceil a \rceil > \lceil x \rceil > \lceil b \rceil$.

4 Let x be a real number. Prove the following.

    a $\lceil \lceil x \rceil \rceil = \lceil x \rceil$

    b $\lceil x \rceil + \lceil y \rceil - 1 \leq \lceil x + y \rceil \leq \lceil x \rceil + \lceil y \rceil$

5 Let x be a real number. Let n and $a_1, \ldots, a_n$ be positive integers. Prove the following.

    a $\left\lceil \dfrac{1}{a_1} \cdot x \right\rceil = \left\lceil \dfrac{x}{a_1} \right\rceil$

    b $\left\lceil \dfrac{1}{a_1} \cdot \left\lceil \dfrac{1}{a_2} \cdot x \right\rceil \right\rceil = \left\lceil \dfrac{x}{a_1 a_2} \right\rceil$

    c $\left\lceil \dfrac{1}{a_1} \cdot \left\lceil \dfrac{1}{a_2} \cdot \left\lceil \dfrac{1}{a_3} \cdot x \right\rceil \right\rceil \right\rceil = \left\lceil \dfrac{x}{a_1 a_2 a_3} \right\rceil$

d $\left\lfloor \frac{1}{a_1} \cdot \left\lceil \frac{1}{a_2} \cdot \left\lfloor \frac{1}{a_3} \cdot \left\lceil \frac{1}{a_4} \cdot x \right\rceil \right\rfloor \right\rceil \right\rfloor = \left\lceil \frac{x}{a_1 a_2 a_3 a_4} \right\rceil$

6.5 Relationship between the floor and ceiling

1 In this section, we prove some theorems which show the relationship between the floor and ceiling functions. The theorems are listed below.

- $\lceil x \rceil = \lfloor x \rfloor$ if and only if x is an integer.
- $\lfloor x \rfloor = \lceil x \rceil - 1$ if and only if x is not an integer.
- $x \leq n \leq y$ if and only if $\lceil x \rceil \leq n \leq \lfloor y \rfloor$.

2 Let x be a real number. Let n be an integer.

3 $\lceil x \rceil = \lfloor x \rfloor$ if and only if x is an integer.

Proof Given x is an integer we have

$$\lfloor x \rfloor = x$$
$$\lceil x \rceil = x$$

Thus,

$$\lfloor x \rfloor = \lceil x \rceil$$

Conversely, we are given

$$\lfloor x \rfloor = \lceil x \rceil \qquad (1)$$

By definition of the floor function

$$\lfloor x \rfloor \leq x$$

Substitute for $\lfloor x \rfloor$ using (1)

$$\lceil x \rceil \leq x \qquad (2)$$

By definition of the ceiling function

$$\lceil x \rceil \geq x$$

From the above inequality and (2), we get

$$\lceil x \rceil = x$$

Due to a previous result the above equality implies x is an integer

■

4   $\lceil x \rceil = \lfloor x \rfloor + 1$ if and only if x is not an integer.

Proof We are given that

$$\lfloor x \rfloor + 1 = \lceil x \rceil$$

Hence

$$\lfloor x \rfloor \neq \lceil x \rceil$$

Due to a previous result we can conclude that x is not an integer. Conversely, we are given that x is not an integer. Thus,

$$\lfloor x \rfloor \neq x$$

Hence,

$$\lfloor x \rfloor + 2 \neq x + 2 \tag{1}$$

We know that

$$\lfloor x \rfloor > x - 1$$

add 2 to both sides

$$\lfloor x \rfloor + 2 > x + 1$$
$$\lfloor x \rfloor + 2 \geq x + 1 \tag{2}$$

By definition of the floor function

$$\lfloor x \rfloor \leq x$$

add 2 to both sides

$$\lfloor x \rfloor + 2 \leq x + 2$$

From (1), we may write the above inequality as

$$\lfloor x \rfloor + 2 < x + 2$$

Chain the above inequality and (2)

$$x + 1 \leq \lfloor x \rfloor + 2 < x + 2$$

By a previous result the above inequality implies

$$\lfloor x \rfloor + 2 = \lceil x + 1 \rceil$$
$$\lfloor x \rfloor + 2 = \lceil x \rceil + 1$$

Subtract 1 from both sides

$$\lfloor x \rfloor + 1 = \lceil x \rceil$$

∎

5 $x \le n \le y$ if and only if $\lceil x \rceil \le n \le \lfloor y \rfloor$.

Proof We are given that $x \le n \le y$. Thus $x \le n$ and $n \le y$. Hence $\lceil x \rceil \le n$ and $n \le \lfloor y \rfloor$. Chaining the last two inequalities, we obtain $\lceil x \rceil \le n \le \lfloor y \rfloor$.

Conversely, we are given that $\lceil x \rceil \le n \le \lfloor y \rfloor$. Thus $\lceil x \rceil \le n$ and $n \le \lfloor y \rfloor$. Hence $x \le n$ and $n \le y$. Chaining the last two inequalities, we obtain $x \le n \le y$. ∎

Exercises 6.5

Let x be a real number. Prove the following.

1 $\lfloor \lceil x \rceil \rfloor = \lceil x \rceil$

2 $\lceil \lfloor x \rfloor \rceil = \lfloor x \rfloor$

3 $x - 1 < \lfloor x \rfloor \le x \le \lceil x \rceil < x + 1$

4 $x + 1 > \lceil x \rceil \ge x \ge \lfloor x \rfloor > x - 1$

6.6 Half of an Integer

1 In this section, we prove the following theorems.

- $\left\lfloor \dfrac{n}{2} \right\rfloor = \dfrac{n}{2}$ if and only if n is even.

- $\left\lceil \dfrac{n}{2} \right\rceil = \dfrac{n}{2}$ if and only if n is even.

- $\left\lfloor \dfrac{n}{2} \right\rfloor = \dfrac{n-1}{2}$ if and only if n is odd.

- $\left\lceil \dfrac{n}{2} \right\rceil = \dfrac{n+1}{2}$ if and only if n is odd.

- $\left\lfloor \dfrac{n}{2} \right\rfloor + \left\lceil \dfrac{n}{2} \right\rceil = n$

2 Let n be an integer.

3 $\left\lfloor \dfrac{n}{2} \right\rfloor = \dfrac{n}{2}$ if and only if n is even.

Proof We are given that n is even. Thus $n/2$ is an integer. Hence $\lfloor n/2 \rfloor = n/2$. Conversely, given that $\lfloor n/2 \rfloor = n/2$ then $n/2$ is an integer. Hence n is even. ∎

4 $\left\lceil \dfrac{n}{2} \right\rceil = \dfrac{n}{2}$ if and only if n is even.

Proof Given that n is even then $n/2$ is an integer. Thus $\lceil n/2 \rceil = n/2$. Conversely, given that $\lceil n/2 \rceil = n/2$ then $n/2$ is an integer. Hence n is even. ∎

5 $\left\lfloor \dfrac{n}{2} \right\rfloor = \dfrac{n-1}{2}$ if and only if n is odd.

Proof

Consider the integers $n, n-1, n-2$. We know that

$$n - 2 < n - 1 < n$$

Thus,

$$n - 2 < n - 1 \leq n$$

Divide both sides by 2

$$\dfrac{n-2}{2} < \dfrac{n-1}{2} \leq \dfrac{n}{2}$$

$$\dfrac{n}{2} - 1 < \dfrac{n-1}{2} \leq \dfrac{n}{2} \tag{1}$$

Given that n is odd then $n-1$ is even. Hence $(n-1)/2$ is an integer. Thus from (1) we can write

$$\left\lfloor \dfrac{n}{2} \right\rfloor = \dfrac{n-1}{2}$$

Conversely, assume

$$\left\lfloor \dfrac{n}{2} \right\rfloor = \dfrac{n-1}{2}$$

Then $(n-1)/2$ is an integer. Hence $n-1$ is even. Consequently, n is odd.

6.6 Half of an Integer

Therefore
$$\left\lfloor \frac{n}{2} \right\rfloor = \frac{n-1}{2}$$
if and only if n is odd.

∎

6 $\left\lceil \dfrac{n}{2} \right\rceil = \dfrac{n+1}{2}$ if and only if n is odd.

Proof Consider the integers $n, n+1, n+2$. We know that
$$n < n+1 < n+2$$
Thus,
$$n \leq n+1 < n+2$$
Divide both sides by 2
$$\frac{n}{2} \leq \frac{n+1}{2} < \frac{n+2}{2}$$
$$\frac{n}{2} \leq \frac{n+1}{2} < \frac{n}{2} + 1 \tag{1}$$

Given that n is odd then $n+1$ is even. Hence $(n+1)/2$ is an integer. Thus from (1) we can write
$$\left\lceil \frac{n}{2} \right\rceil = \frac{n+1}{2}$$

Conversely, assume
$$\left\lceil \frac{n}{2} \right\rceil = \frac{n+1}{2}$$
Then $(n+1)/2$ is an integer. Hence $n+1$ is even. Consequently, n is odd. Therefore
$$\left\lceil \frac{n}{2} \right\rceil = \frac{n+1}{2}$$
if and only if n is odd.

∎

7 $\left\lfloor \dfrac{n}{2} \right\rfloor + \left\lceil \dfrac{n}{2} \right\rceil = n$

Proof When n is even we have that
$$\left\lfloor \frac{n}{2} \right\rfloor = \frac{n}{2}$$
and
$$\left\lceil \frac{n}{2} \right\rceil = \frac{n}{2}$$
add both equations
$$\left\lfloor \frac{n}{2} \right\rfloor + \left\lceil \frac{n}{2} \right\rceil = n$$

When n is odd we have that
$$\left\lfloor \frac{n}{2} \right\rfloor = \frac{n-1}{2}$$
and
$$\left\lceil \frac{n}{2} \right\rceil = \frac{n+1}{2}$$
add both equations
$$\begin{aligned} \left\lfloor \frac{n}{2} \right\rfloor + \left\lceil \frac{n}{2} \right\rceil &= \frac{n-1}{2} + \frac{n+1}{2} \\ &= \frac{n-1+n+1}{2} \\ &= \frac{2n}{2} \\ &= n \end{aligned}$$

Therefore in both cases we have that
$$\left\lfloor \frac{n}{2} \right\rfloor + \left\lceil \frac{n}{2} \right\rceil = n$$

■

CHAPTER 7

Roots

7.1 Square Root

1. In this section, we define square root and provide some examples. Then we prove that all positive real numbers have exactly two square roots.

2. **Definition of square root.** A square root of a real number y is a real number x such that $y = x^2$.

3. For example, we know that

$$4 = (-2)^2 \quad \text{and} \quad 4 = 2^2$$

   Hence the square roots of 4 are -2 and 2.

4. **Zero and Square Root.** Let x, y be real numbers.

5. $x^2 \geq 0$.

   Proof It is either $x \geq 0$ or $x < 0$.

   When $x \geq 0$, we can square both sides of the inequality. This gives $x^2 \geq 0^2$. Hence $x^2 \geq 0$.

   When $x < 0$, we multiply both sides by -1. This gives $-x > 0$. Squaring both sides, we obtain $x^2 > 0$. Thus $x^2 \geq 0$. Therefore in both cases $x^2 \geq 0$. ∎

6. Zero is a square root of y if and only if y is zero.

   Proof Given that zero is a square root of y, then $y = 0^2$. Hence $y = 0$. Conversely, given $y = 0$ then $y = 0^2$. Hence zero is a square root of y. ∎

7. If y has a square root then $y \geq 0$.

Proof Given that y has a square root then

$$y = x^2 \quad \text{for some real number } x \qquad (1)$$

Thus,

$$x^2 \geq 0$$

Use (1) to substitute for x^2

$$y \geq 0$$

∎

8   If $y < 0$ then y does not have a square root.

Proof The statement is the contrapositive of the previous result. Hence it is true. ∎

9   If $y \geq 0$ then y has a square root

(Existence of Square Root)

Note The proof of this result requires methods we have not yet learnt. We shall take the result as an axiom.

10   If $y > 0$ then all square roots of y are nonzero.

Proof Since $y > 0$, y has a square root. And since $y \neq 0$, zero is not a root of y. Hence all square roots of y are nonzero. ∎

11   **Exactly Two Square Roots.** Let y be a nonnegative real number and let $y = x^2$ for some real number x.

12   $|x|$ is a square root of y

Proof Observe that $x^2 = |x|^2$. Thus $y = |x|^2$. Hence $|x|$ is a square root of y. ∎

13   $-|x|$ is a square root of y

Proof

Observe that

$$(-|x|)^2 = |x|^2$$
$$= x^2$$
$$= y$$

Thus,

$$y = (-|x|)^2$$

Hence $-|x|$ is a square root of y ∎

14 $|x|$ and $-|x|$ are the only square roots of y

Proof Let r be a square root of y. Thus

$$y = r^2$$

We are given that

$$y = x^2$$

Thus,

$$x^2 = r^2$$
$$|x| = |r|$$
$$r = \pm |x|$$
$$r = |x| \quad \text{or} \quad r = -|x|$$

Thus $|x|$ and $|-x|$ are the only square roots of y. ∎

15 Every nonnegative real number has a unique nonnegative square root.

Proof Let y be a nonnegative real number. Thus $y = x^2$ for some real number x.

If $y = 0$ then the only square root is zero. Hence y has a unique nonnegative square root.

If $y > 0$ then $-|x|$ and $|x|$ are the only square roots of y. Since y is nonzero then all its square roots are nonzero. Thus $|x| \neq 0$. Hence $|x| \geq 0$ and $|x| \neq 0$. Thus $|x| > 0$ and $-|x| < 0$. Hence y has exactly one positive and one negative square root. Hence the positive square root is the only nonnegative square root of y. Thus y has a unique nonnegative square root.

In both cases y has a unique nonnegative square root. ∎

16 The principal square root of a nonnegative real number is the nonnegative square root of the number.

(Definition of Principal Square Root)

17 The principal square root of a nonnegative real number x is denoted by

$$\sqrt{x}$$

The principal square root is also simply called *the square root*.

(Notation of Principal Square Root)

7.2 Properties of the Square Root

1 In this section, we prove the following properties of square roots.

- $\sqrt{x^2} = |x|$
- If $y \geq 0$ then $(\sqrt{y})^2 = y$
- If $x, y \geq 0$ then $\sqrt{xy} = \sqrt{x}\sqrt{y}$

2 Let x, y be real numbers.

3 If $y \geq 0$ then $\sqrt{y} = |x|$ if and only if $y = x^2$.

Proof Given $\sqrt{y} = |x|$ then $|x|$ is a square root of y. Hence $|x|^2 = y$. Thus $x^2 = y$. Conversely given $y = x^2$, the nonnegative square root of y is $|x|$. Hence $\sqrt{y} = |x|$. ∎

4 $\sqrt{x^2} = |x|$

Proof Let
$$y = x^2 \tag{1}$$
Thus
$$\sqrt{y} = |x|$$
Use (1) to substitute for y in the above equation
$$\sqrt{x^2} = |x|$$

∎

5. If $y \geq 0$ then $(\sqrt{y})^2 = y$

   Proof Let
   $$y = x^2 \tag{1}$$
   Hence
   $$\sqrt{y} = |x| \tag{2}$$
   From (1) we may write
   $$y = |x|^2$$
   If we substitute for $|x|$ using (2) we get
   $$y = (\sqrt{y})^2$$

   ∎

6. If $x, y \geq 0$ then $\sqrt{xy} = \sqrt{x}\sqrt{y}$

   Proof Let
   $$w = \sqrt{x} \tag{1}$$
   $$z = \sqrt{y} \tag{2}$$
   Square both equations
   $$w^2 = x$$

$$z^2 = y$$

Multiply both equations together

$$w^2 \cdot z^2 = xy$$
$$(wz)^2 = xy$$

Hence

$$\sqrt{xy} = |wz|$$

Since w and z are principal square roots, they are nonnegative. Hence

$$\sqrt{xy} = wz$$

Using (1) and (2) we can substitute for w and z

$$\sqrt{xy} = \sqrt{x}\sqrt{y}$$

∎

Exercises 7.2

Let x, y, z be real numbers. Let n be a positive integer and let $x_1, x_2, \ldots, x_n$ be real numbers. Prove the following.

1. If $x, y, z \geq 0$ then $\sqrt{xyz} = \sqrt{x}\sqrt{y}\sqrt{z}$

2. If $x_1, \ldots, x_n \geq 0$ then $\sqrt{x_1 \cdots x_n} = \sqrt{x_1} \cdots \sqrt{x_n}$

3. $\sqrt{\sqrt{x^4}} = |x|$

7.3 Cube Root

1. In this section, we define cube root and provide some examples. Then we prove that every real number has a unique cube root.

2. **Definition of cube root.** A cube root of a real number y is a real number x such that $y = x^3$.

3. For example, we know that
$$8 = 2^3$$

Hence the cube root of 8 is 2.

7.3 Cube Root

4 Zero is a cube root of a real number if and only if the real number is zero.

   Proof Let y be a real number with cube root of zero. Thus $y = 0^3$. Hence $y = 0$. Conversely, given $y = 0$ then $y = 0^3$. Hence zero is a cube root of y. ∎

5 Every real number has a cube root.

   (Existence of Cube Root)

   Note The proof of this result requires methods we have not yet learnt. We shall take the result as an axiom.

6 **Uniqueness of cube root.** Here we prove that each real number has a unique cube root. Here is a list of the theorems we prove:

   - If $w^3 = x^3$ then $w = x$.
   - Every real number has a unique cube root.

7 Let w, x be real numbers.

8 $x \geq 0$ if and only if $x^3 \geq 0$.

   Proof Let $x \geq 0$. Thus $\operatorname{sgn} x = 0$ or $\operatorname{sgn} x = 1$. Since $\operatorname{sgn} x^3 = \operatorname{sgn} x$, it follows that $\operatorname{sgn} x^3 = 0$ or $\operatorname{sgn} x^3 = 1$. Thus $x^3 \geq 0$.

   Conversely, let $x^3 \geq 0$. Thus $\operatorname{sgn} x^3 = 0$ or $\operatorname{sgn} x^3 = 1$. Since $\operatorname{sgn} x = \operatorname{sgn} x^3$, it follows that $\operatorname{sgn} x = 0$ or $\operatorname{sgn} x = 1$. Thus $x \geq 0$. ∎

9 $x < 0$ if and only if $x^3 < 0$.

   Proof This is the contrapositive of the previous result. ∎

10 If $w^3 = x^3$ then $w = x$.

   Proof We are given that $w^3 = x^3$. Take absolute value of both sides. Thus $|w^3| = |x^3|$. Hence $|w| = |x|$. It is either $w^3 \geq 0$ or $w^3 < 0$. We shall consider each case.

   When $w^3 \geq 0$, we have that $x^3 \geq 0$. Consequently, $w, x \geq 0$. Thus $|w| = w$ and $|x| = x$. Since $|w| = |x|$, it follows that $w = x$.

   When $w^3 < 0$, we have that $x^3 < 0$. Consequently, $w, x < 0$. Thus $|w| = -w$ and $|x| = -x$. Since $|w| = |x|$, it follows that $-w = -x$. Therefore $w = x$.

   In both cases $w = x$. ∎

11 Every real number has a unique cube root.

   Proof First, we prove existence. Let y be a real number. From a previous result, y has a cube root.

   Second, we prove uniqueness. Let w, x be cube roots of y. Hence $y = w^3$ and $y = x^3$. Thus $w^3 = x^3$. Consequently, $w = x$. Therefore the cube root of y is unique. Hence every real number has a unique cube root. ∎

12 The cube root of a nonnegative real number x is denoted by
$$\sqrt[3]{x}.$$

7.4 Properties of the Cube Root

1 In this section, we prove the following properties of cube root.

   - $\sqrt[3]{y} = x$ if and only if $y = x^3$.
   - $\sqrt[3]{x^3} = x$
   - $(\sqrt[3]{y})^3 = y$
   - $\sqrt[3]{xy} = \sqrt[3]{x}\sqrt[3]{y}$

2 Let x, y be real numbers.

3 $\sqrt[3]{y} = x$ if and only if $y = x^3$.

   Proof We are given that $\sqrt[3]{y} = x$. Thus x is the cube root of y. Hence $y = x^3$. Conversely, if $y = x^3$, then the cube root of y is x. Hence $\sqrt[3]{y} = x$. ∎

4 $\sqrt[3]{x^3} = x$

   Proof Let
   $$y = x^3 \tag{1}$$

   Thus
   $$\sqrt[3]{y} = x$$

   Use (1) to substitute for y in the above equation
   $$\sqrt[3]{x^3} = x$$

   ∎

7.4 Properties of the Cube Root

5 $(\sqrt[3]{y})^3 = y$

Proof Let

$$y = x^3 \qquad (1)$$

Hence

$$\sqrt[3]{y} = x \qquad (2)$$

Cube both sides

$$(\sqrt[3]{y})^3 = x^3$$

From (1) we may substitute for x^3 in the above equation

$$(\sqrt[3]{y})^3 = y$$

∎

6 $\sqrt[3]{xy} = \sqrt[3]{x}\sqrt[3]{y}$

Proof Let

$$w = \sqrt[3]{x} \qquad (1)$$
$$z = \sqrt[3]{y} \qquad (2)$$

Cube both sides of each equation

$$w^3 = x$$
$$z^3 = y$$

Multiply both equations together

$$w^3 \cdot z^3 = xy$$
$$(wz)^3 = xy$$

Hence

$$\sqrt[3]{xy} = wz$$

Using (1) and (2) we can substitute for w and z

$$\sqrt[3]{xy} = \sqrt[3]{x}\sqrt[3]{y}$$

∎

Exercises 7.4

Let x, y, z be real numbers. Let n be a positive integer and let $x_1, x_2, \ldots, x_n$ be real numbers. Prove the following.

1 $\sqrt[3]{xyz} = \sqrt[3]{x}\sqrt[3]{y}\sqrt[3]{z}$

2 $\sqrt[3]{x_1 \cdots x_n} = \sqrt[3]{x_1} \cdots \sqrt[3]{x_n}$

3 $\sqrt[3]{\sqrt[3]{x^9}} = x$

7.5 Nth Root

1 In this section, we define nth root and provide some examples. Then we prove that every nonnegative real number has a unique nonnegative nth root.

2 **Definition of nth root.** Let y be a nonnegative real number.

- a square root of y is a real number x, such that $y = x^2$
- a cube root of y is a real number x, such that $y = x^3$
- a fourth root of y is a real number x, such that $y = x^4$
- for all integers $n > 1$, an nth root of y is a real number x, such that $y = x^n$

3 **Uniqueness of the nonnegative nth root.** Here, we prove that every nonnegative real number has a unique nonnegative nth root. Here is a list of some properties we prove.

- For each integer $n > 1$, every nonnegative real number has a nonnegative nth root.
  (Existence of the Nonnegative nth Root)
- If $w, x \geq 0$ and $w^n = x^n$ then $w = x$.
  (Uniqueness of the Nonnegative nth Root)
- For each integer $n > 1$, every nonnegative real number has a unique nonnegative nth root.

4 For each integer $n > 1$, every nonnegative real number has an nth root.

(Existence of nth Root)

7.5 Nth Root

Note The proof of this result requires methods we have not yet learnt. We shall take the result as an axiom.

5 For each integer $n > 1$, every nonnegative real number has a nonnegative nth root.

(Existence of the Nonnegative nth Root)

Proof

Let y be a nonnegative real number. Thus y has an nth root, say x. Thus $y = x^n$. Find absolute value of both sides. Hence $|y| = |x^n|$. Since y is nonnegative $|y| = y$. Thus $y = |x^n|$. It follows that $y = |x|^n$. Hence $|x|$ is an nth root of y. Since absolute values are always nonnegative, $|x|$ is nonnegative. Therefore y has a nonnegative nth root. ∎

6 If $w, x \geq 0$ and $w^n = x^n$ then $w = x$.

(Uniqueness of the Nonnegative nth Root)

Proof We are given that $w^n = x^n$. Find absolute value of both sides. Thus $|w^n| = |x^n|$. Hence $|w| = |x|$. Since $w, x \geq 0$, it follows that $|w| = w$ and $|x| = x$. Thus $w = x$. ∎

7 For each integer $n > 1$, every nonnegative real number has a unique nonnegative nth root.

Proof First, we prove existence. Let y be a nonnegative real number. Thus y has a nonnegative nth root.

Second, we prove uniqueness. Let w, x be nonnegative nth roots of y. Hence $y = w^n$ and $y = x^n$. Thus $w^n = x^n$. Consequently $w = x$. Therefore the nonnegative nth root of y is unique. Hence every real number has a unique nth root. ∎

8 For all nonnegative integer n, the nth root of y is the nonnegative real number x, such that $y = x^n$. It is denoted by $\sqrt[n]{y}$.

(Definition of the Nth Root)

7.6 Properties of the Nth Root

1. In this section, we prove the following properties of the nth root.

   - $\sqrt[n]{y} = x$ if and only if $y = x^n$.
   - $\sqrt[n]{x^n} = x$
   - $(\sqrt[n]{y})^n = y$
   - $(\sqrt[n]{x})^n = \sqrt[n]{x^n}$
   - $\sqrt[m]{\sqrt[n]{x}} = \sqrt[mn]{x}$
   - $\sqrt[n]{xy} = \sqrt[n]{x}\sqrt[n]{y}$
   - $x > y$ if and only if $\sqrt[n]{x} > \sqrt[n]{y}$ (Root of Inequality)

2. Let x, y be nonnegative real numbers and let n be an integer greater than 1.

3. $\sqrt[n]{y} = x$ if and only if $y = x^n$.

   Proof We are given that $\sqrt[n]{y} = x$. Thus x is the nth root of y. Hence $y = x^n$. Conversely, if $y = x^n$, then the nth root of y is x. Hence $\sqrt[n]{y} = x$. ∎

4. $\sqrt[n]{x^n} = x$

   Proof Let
   $$y = x^n \qquad (1)$$
   Thus
   $$\sqrt[n]{y} = x$$
   Use (1) to substitute for y in the above equation
   $$\sqrt[n]{x^n} = x$$
   ∎

5. $(\sqrt[n]{y})^n = y$

7.6 Properties of the Nth Root

Proof Let

$$y = x^n \qquad (1)$$

Hence

$$\sqrt[n]{y} = x \qquad (2)$$

Raise both sides to the nth power

$$(\sqrt[n]{y})^n = x^n$$

From (1) we may substitute for x^n in the above equation

$$(\sqrt[n]{y})^n = y$$

∎

6. $(\sqrt[n]{x})^n = \sqrt[n]{x^n}$

Proof

From previous results $(\sqrt[n]{x})^n = x$ and $\sqrt[n]{x^n} = x$. Due to transitivity, $(\sqrt[n]{x})^n = \sqrt[n]{x^n}$. ∎

7. $\sqrt[m]{\sqrt[n]{x}} = \sqrt[mn]{x}$

Proof Let z be the (mn)th root of x. That is, $z^{mn} = x$. Thus $(z^m)^n = x$. Hence $z^m = \sqrt[n]{x}$. It follows that $z = \sqrt[m]{\sqrt[n]{x}}$. Therefore the (mn)th root of x is $\sqrt[m]{\sqrt[n]{x}}$. Hence $\sqrt[m]{\sqrt[n]{x}} = \sqrt[mn]{x}$. ∎

8. $\sqrt[n]{xy} = \sqrt[n]{x}\sqrt[n]{y}$

Proof Let

$$w = \sqrt[n]{x} \qquad (1)$$
$$z = \sqrt[n]{y} \qquad (2)$$

Raise both sides to the nth power

$$w^n = x$$

$$z^n = y$$

Multiply both equations together

$$w^n \cdot z^n = xy$$
$$(wz)^n = xy$$

Hence

$$\sqrt[n]{xy} = wz$$

Using (1) and (2) we can substitute for w and z

$$\sqrt[n]{xy} = \sqrt[n]{x}\sqrt[n]{y}$$

∎

9. If $x, y \geq 0$ and $x > y$ then $\sqrt[n]{x} > \sqrt[n]{y}$.

Proof We are given that $x, y \geq 0$. Thus x, y have unique nonnegative nth roots. Let $w = \sqrt[n]{x}$ and $z = \sqrt[n]{y}$. Thus $w^n = x$ and $z^n = y$. Hence we can write $x > y$ as $w^n > z^n$. Since $w, z \geq 0$ and $w^n > z^n$, it follows that $w > z$. Therefore $\sqrt[n]{x} > \sqrt[n]{y}$. ∎

10. If $x, y \geq 0$ and $\sqrt[n]{x} > \sqrt[n]{y}$ then $x > y$.

Proof We are given that

$$\sqrt[n]{x} > \sqrt[n]{y}$$

Raise both sides to the nth power

$$(\sqrt[n]{x})^n > (\sqrt[n]{y})^n$$
$$x > y$$

∎

11. If $x, y \geq 0$ then $x > y$ if and only if $\sqrt[n]{x} > \sqrt[n]{y}$.

Proof This is the conjunction of the previous two results. ∎

Exercises 7.6

1 Let a, b, x be real numbers. Prove the following.

a If $x, y \geq 0$ and $x < y$ then $\sqrt[n]{x} < \sqrt[n]{y}$.

b If $x, y \geq 0$ and $x \geq y$ then $\sqrt[n]{x} \geq \sqrt[n]{y}$.

c If $x, y \geq 0$ and $x \leq y$ then $\sqrt[n]{x} \leq \sqrt[n]{y}$.

2 Let a, b, x be real numbers. Prove the following.

a If $a \geq 0$ and $a \leq x \leq b$ then $\sqrt[n]{a} \leq \sqrt[n]{x} \leq \sqrt[n]{b}$.

b If $b \geq 0$ and $a \geq x \geq b$ then $\sqrt[n]{a} \geq \sqrt[n]{x} \geq \sqrt[n]{b}$.

3 Let a, b, x be real numbers. Prove the following.

a If $a \geq 0$ and $a < x < b$ then $\sqrt[n]{a} < \sqrt[n]{x} < \sqrt[n]{b}$.

b If $b \geq 0$ and $a > x > b$ then $\sqrt[n]{a} > \sqrt[n]{x} > \sqrt[n]{b}$.

4 Let x, y, z be nonnegative real numbers. Let n, m be positive integers and let $x_1, x_2, \ldots, x_m$ be nonnegative real numbers. Prove the following.

a If $n > 1$ then $\sqrt[n]{xyz} = \sqrt[n]{x} \sqrt[n]{y} \sqrt[n]{z}$.

b If $n > 1$ then $\sqrt[n]{x_1 \cdots x_m} = \sqrt[n]{x_1} \cdots \sqrt[n]{x_m}$.

c $\sqrt[n]{\sqrt[n]{x^{n^2}}} = x$

CHAPTER 8

Rational Exponents

8.1 Power of Power Law

1. We have learnt about integer exponents: x^1, x^{-2}, etc. Now we extend the idea of exponents to rational numbers: $x^{1/2}$, $x^{-\frac{2}{3}}$, etc. In order to work with rational exponents, we need three theorems:

   - power of power law
   - different bases law
   - same base law

2. In this section, we prove the power of power law: for all positive real numbers x and for all rational numbers r, s,
$$(x^r)^s = x^{r \cdot s}.$$

   To achieve this, we first consider the case where r, s are positive integers. Then, we define negative rational number exponents and prove some theorems that enable us to work with such exponents. Finally, we use all the theorems we have proven thus far to prove the power of power law.

3. **Positive rational number exponents.** Let x be a positive real number. Let m, n, p, q be positive integers and let r, s be rational numbers.

4. Let us define $x^{1/n} = \sqrt[n]{x}$ for all integers $n > 1$.

5. $(x^{1/n})^n = (x^n)^{1/n}$

   Proof

   $$(\sqrt[n]{x})^n = \sqrt[n]{x^n}$$

We can rewrite the above equation using exponentiation notation

$$(x^{1/n})^n = (x^n)^{1/n}$$

∎

6 $(x^{1/n})^{1/m} = x^{1/mn}$

Proof From a previous result, $\sqrt[m]{\sqrt[n]{x}} = \sqrt[mn]{x}$. If we rewrite this result using exponentiation notation, we get $(x^{1/n})^{1/m} = x^{1/mn}$. ∎

7 $(x^{1/n})^{1/m} = (x^{1/m})^{1/n}$

Proof

$$(x^{1/n})^{1/m} = x^{1/mn}$$
$$= (x^{1/m})^{1/n}$$

∎

8 $(x^{1/n})^m = (x^m)^{1/n}$

Proof Let z be the nth root of x^m.

$$z^n = x^m \tag{1}$$

Thus

$$z = (x^m)^{1/n} \tag{2}$$

Also, from (1),

$$x = (z^n)^{1/m}$$

Find nth root of both sides

$$x^{1/n} = [(z^n)^{1/m}]^{1/n}$$
$$= [(z^n)^{1/n}]^{1/m}$$
$$= z^{1/m}$$

Raise both sides to the power m

$$(x^{1/n})^m = (z^{1/m})^m$$
$$= z$$

From (2) we can substitute for z

$$(x^{1/n})^m = (x^m)^{1/n}$$

∎

9 Let us define $x^{m/n} = (x^{1/n})^m$ for all integers $n, m > 1$.

10 $(x^{p/q})^{m/n} = x^{pm/qn}$

Proof Let z be the (qn)th root of x^{pm}. That is, $z = x^{pm/qn}$. Thus,

$$z^{qn} = x^{pm}$$
$$(z^n)^q = x^{pm}$$

Find qth root of both sides

$$z^n = (x^{pm})^{1/q}$$
$$= [(x^p)^m]^{1/q}$$
$$= [(x^p)^{1/q}]^m$$
$$= (x^{p/q})^m$$

Find nth root of both sides

$$z = [(x^{p/q})^m]^{1/n}$$
$$= (x^{p/q})^{m/n}$$

Since $z = x^{pm/qn}$, we may write

$$x^{pm/qn} = (x^{p/q})^{m/n}$$

∎

11 If $r, s \geq 0$ then $(x^r)^s = x^{rs}$

Proof Consider the rational numbers r, s. It is either one among them is zero or none of them is zero. Hence $r = 0$ or $s = 0$ or $r, s \neq 0$. Thus $r = 0$ or $s = 0$ or $r, s > 0$.

When $r = 0$ we get $x^r = 1$. Thus $(x^r)^s = 1^s$. Hence $(x^r)^s = 1$. Also, $x^{rs} = x^0$. Hence $x^{rs} = 1$. Therefore $(x^r)^s = x^{rs}$.

When $s = 0$ we get $(x^r)^s = (x^r)^0$. Thus $(x^r)^s = 1$. Also, $x^{rs} = x^0$. Hence $x^{rs} = 1$. Therefore $(x^r)^s = x^{rs}$.

Consider the case $r, s > 0$. We can write $r = p/q$ and $s = m/n$ for some positive integers m, n, p, q. Thus

$$(x^r)^s = (x^{p/q})^{m/n}$$
$$= x^{pq/mn}$$
$$= x^{rs}$$

∎

12 **Negative rational number exponents.** Now we discuss negative rational number exponents. Let x be a positive real number.

13 $(x^{-1})^{1/n} = (x^{1/n})^{-1}$

Proof Let z be the nth root of x^{-1}. Thus

$$z^n = x^{-1} \tag{1}$$

Find the nth root of both sides

$$z = (x^{-1})^{1/n} \tag{2}$$

From (1)

$$(z^n)^{-1} = x$$
$$(z^{-1})^n = x$$

Find the nth root of both sides

$$z^{-1} = x^{1/n}$$
$$z = (x^{1/n})^{-1}$$

From (2) we can substitute for z

$$(x^{-1})^{1/n} = (x^{1/n})^{-1}$$

∎

14 Let us define $x^{-1/n} = (x^{1/n})^{-1}$ for all integers $n > 1$.

15 $(x^{-1})^{m/n} = (x^{m/n})^{-1}$

Proof

$$\begin{aligned}(x^{-1})^{m/n} &= [(x^{-1})^m]^{1/n} \\ &= [(x^m)^{-1}]^{1/n} \\ &= [(x^m)^{1/n}]^{-1} \\ &= (x^{m/n})^{-1}\end{aligned}$$

∎

16 Let us define $x^{-m/n} = (x^{m/n})^{-1}$ for all integers $n, m > 1$.

17 **Power of power law.** The theorems we proved above are used to prove one theorem which unifies them: the power of power law. It states that $(x^r)^s = x^{rs}$ for all rational numbers r, s and for all positive real numbers x. Below we prove this theorem by proving four cases.

18 Let x be a positive real number and let r, s be rational numbers.

19 $(x^r)^s = x^{rs}$

Proof There are four cases:

1. $r \geq 0$ and $s \geq 0$
2. $r \geq 0$ and $s \leq 0$
3. $r \leq 0$ and $s \geq 0$
4. $r \leq 0$ and $s \leq 0$

In the first case, a previous result gives $(x^r)^s = x^{rs}$.

In the second case, r and $-s$ are nonnegative rational numbers. Thus

$$\begin{aligned}(x^r)^{-s} &= x^{-rs} \\ [(x^r)^s]^{-1} &= (x^{rs})^{-1} \\ (x^r)^s &= x^{rs}\end{aligned}$$

In the third case, $-r$ and s are nonnegative rational numbers. Thus

$$(x^{-r})^s = x^{-rs}$$

$$[(x^r)^{-1}]^s = (x^{rs})^{-1}$$
$$[(x^r)^s]^{-1} = (x^{rs})^{-1}$$
$$(x^r)^s = x^{rs}$$

In the fourth case, $-r$ and $-s$ are nonnegative rational numbers. Thus

$$(x^{-r})^{-s} = x^{(-r)(-s)}$$
$$[(x^r)^{-1}]^{-s} = (x^{(-r)s})^{-1}$$
$$[(x^r)^{-s}]^{-1} = (x^{(-r)s})^{-1}$$
$$(x^r)^{-s} = x^{(-r)s}$$
$$[(x^r)^s]^{-1} = (x^{rs})^{-1}$$
$$(x^r)^s = x^{rs}$$

In conclusion, we get $(x^r)^s = x^{rs}$ in all four cases. ∎

Exercises 8.1

Let x be a positive real number and let a, b, c, d be rational numbers. Prove the following.

1 $((x^a)^b)^c = x^{abc}$

2 $(((x^a)^b)^c)^d = x^{abcd}$

8.2 Different Bases Law

1 We prove that, for all rational numbers r and for all positive real numbers x, y,

$$x^r y^r = (xy)^r.$$

We call this theorem the different bases law. First, we prove the case where the exponent, r, is positive. Next, we prove the case where the exponent, r, is negative. By combining these two cases, we prove the different bases law.

2 **Positive rational number exponent.** Let x, y be positive real numbers and let m, n be positive integers.

3 $x^{1/n} y^{1/n} = (xy)^{1/n}$

Proof From a previous result $\sqrt[n]{x} \sqrt[n]{y} = \sqrt[n]{xy}$. We can rewrite the previous equation using exponentiation notation: $x^{1/n} y^{1/n} = (xy)^{1/n}$. ∎

4    $x^{m/n} y^{m/n} = (xy)^{m/n}$

Proof

$$\begin{aligned} x^{m/n} y^{m/n} &= (x^m)^{1/n} (y^m)^{1/n} \\ &= (x^m y^m)^{1/n} \\ &= [(xy)^m]^{1/n} \\ &= (xy)^{m/n} \end{aligned}$$

■

5   **Negative rational number exponent.** Let x, y be positive real numbers and let m, n be positive integers.

6    $x^{-m/n} y^{-m/n} = (xy)^{-m/n}$

Proof

$$\begin{aligned} x^{-m/n} y^{-m/n} &= (x^{m/n})^{-1} (y^{m/n})^{-1} \\ &= (x^{m/n} y^{m/n})^{-1} \\ &= [(xy)^{m/n}]^{-1} \\ &= (xy)^{-m/n} \end{aligned}$$

■

7   **Different bases law.** Let x, y be positive real numbers. Let m, n be positive integers and let r be a rational number.

8    $x^r y^r = (xy)^r$

Proof It is either $r = 0$ or $r > 0$ or $r < 0$. When $r = 0$ we have $x^r y^r = x^0 y^0$. Thus $x^r y^r = 1$. Also, $(xy)^r = (xy)^0$. Thus $(xy)^r = 1$. Therefore $x^r y^r = (xy)^r$. When $r > 0$ we can write $r = m/n$ for some positive integers m, n. Thus $x^r y^r = x^{m/n} y^{m/n}$. From a previous result, $x^{m/n} y^{m/n} = (xy)^{m/n}$. Therefore $x^r y^r = (xy)^r$.

When $r < 0$ we can write $r = -m/n$ for some positive integers m, n. Thus $x^r y^r = x^{-m/n} y^{-m/n}$. From a previous result, $x^{-m/n} y^{-m/n} = (xy)^{-m/n}$. Therefore $x^r y^r = (xy)^r$.

In all cases, $x^r y^r = (xy)^r$.

■

Exercises 8.2

Let x, y, z be a positive real number. Let r be a rational number. Let n be a positive integer and let $x_1, \ldots, x_n$ be positive real numbers. Prove the following.

1 $x^r \cdot y^r \cdot z^r = (xyz)^z$

2 $x_1^r \cdots x_n^r = (x_1 \cdots x_n)^r$

8.3 Same Base Law

1 In this section, we prove that, for all rational numbers r, s and for all positive real numbers x,
$$x^r x^s = x^{r+s}.$$

We call this theorem the same base law. We prove the case where both exponents are nonnegative. Then we prove the case where one exponent is nonnegative and the other is not positive. Additionally, we prove the case when both exponents are not positive. We combine these cases to prove the same base law.

2 **Both exponents are nonnegative.** Let x be a positive real number. Let m, n, p, q be positive integers and let r, s be rational numbers.

3 $x^{1/n} x^{1/m} = x^{1/n + 1/m}$

Proof

$$x^m x^n = x^{m+n}$$

Find the mnth root of both sides

$$(x^m x^n)^{1/mn} = (x^{m+n})^{1/mn}$$
$$(x^m)^{1/mn} (x^n)^{1/mn} = (x^{(m+n)/mn})$$
$$x^{m/mn} x^{n/mn} = x^{m/mn + n/mn}$$
$$x^{1/n} x^{1/m} = x^{1/n + 1/m}$$

∎

4 $x^{m/n} x^{p/q} = x^{m/n + p/q}$

Proof

$$x^{mq}x^{np} = x^{mq+np}$$

Find the qnth root of both sides

$$(x^{mq}x^{np})^{1/qn} = (x^{mq+np})^{1/qn}$$
$$(x^{mq})^{1/qn}(x^{np})^{1/qn} = x^{(mq+np)/qn}$$
$$x^{mq/qn}x^{np/qn} = x^{mq/qn+np/qn}$$
$$x^{m/n}x^{p/q} = x^{m/n+p/q}$$

∎

5. If $r, s \geq 0$ then $x^r x^s = x^{r+s}$.

Proof Consider the nonnegative integers r, s. It is either one among them is zero or none of them is zero. Hence $r = 0$ or $s = 0$ or $r, s \neq 0$. Thus $r = 0$ or $s = 0$ or $r, s > 0$.

When $r = 0$, we have $x^r x^s = x^0 x^s$. Thus $x^r x^s = x^s$. Hence $x^r x^s = x^{0+s}$. That is, $x^r x^s = x^{r+s}$.

When $s = 0$, we have $x^r x^s = x^r x^0$. Thus $x^r x^s = x^r$. Hence $x^r x^s = x^{r+0}$. That is, $x^r x^s = x^{r+s}$.

When $r, s > 0$, we can write $r = m/n$ and $s = p/q$ for some positive integers m, n, p, q. By a previous result, $x^{m/n} x^{p/q} = x^{m/n+p/q}$. Thus $x^r x^s = x^{r+s}$.

Therefore, in all cases $x^r x^s = x^{r+s}$. ∎

6. **One exponent is nonnegative and the other is not positive.** Let x be a positive real number. For all nonnegative rational numbers r, s,

$$x^r x^{-s} = x^{r-s}.$$

We prove this theorem for positive $r - s$, for negative $r - s$, and for $r - s = 0$. Then we conclude that $x^r x^{-s} = x^{r-s}$ for all nonnegative rational numbers r, s.

7. If $r, s \geq 0$ and $r - s$ is nonnegative then $x^r x^{-s} = x^{r-s}$.

Proof

We are given that $r - s$ is a nonnegative rational number. Thus

$$x^{r-s} = x^{r-s} \cdot 1$$

$$= x^{r-s}x^s x^{-s}$$
$$= x^{r-s+s}x^{-s}$$
$$= x^r x^{-s}$$

8. If $r, s \geq 0$ and $r - s$ is negative then $x^r x^{-s} = x^{r-s}$.

   Proof

   We are given that $r - s$ is a negative rational number. Thus $s - r$ is a positive rational number. Hence
   $$x^s x^{-r} = x^{s-r}$$

   Find the reciprocal of both sides
   $$(x^s x^{-r})^{-1} = (x^{s-r})^{-1}$$
   $$(x^s)^{-1}(x^{-r})^{-1} = (x^{s-r})^{-1}$$
   $$x^{-s} x^r = x^{-(s-r)}$$
   $$x^r x^{-s} = x^{r-s}$$

9. If $r, s \geq 0$ then $x^r x^{-s} = x^{r-s}$.

   Proof It is either $r - s$ is nonnegative or $r - s$ is negative. In each case, a previous result gives $x^r x^{-s} = x^{r-s}$.

10. **Both exponents are not positive.** Let x be a positive real number. For all nonnegative rational numbers r, s,
    $$x^{-r} x^{-s} = x^{-r-s}.$$

    Proof
    $$x^{-r} x^{-s} = (x^r)^{-1}(x^s)^{-1}$$
    $$= (x^r x^s)^{-1}$$
    $$= (x^{r+s})^{-1}$$
    $$= x^{-(r+s)}$$
    $$= x^{-r-s}$$

11 **Same base law.** We unify all previous theorems in this section by using them to prove the same base law. This theorem states that $x^r x^s = x^{r+s}$ for all rational numbers r, s and for all positive real numbers x.

12 Let x be a positive real number and let r, s be rational numbers.

13 $x^r x^s = x^{r+s}$

  Proof There are three cases:

  1. both exponents are nonnegative
  2. both exponents are not positive
  3. one exponent is nonnegative while the other is not positive

  Each of these cases is proved by a previous result. Thus, $x^r x^s = x^{r+s}$ for all positive real numbers x and for all rational numbers r, s. ∎

Exercises 8.3

Let x be a positive real number. Let a, b, c be rational numbers. Let n be a positive integer and let $a_1, \ldots, a_n$ be rational numbers. Prove the following.

1 $x^a x^b x^c = x^{a+b+c}$

2 $x^{a_1} \cdots x^{a_n} = x^{a_1 + \cdots + a_n}$

CHAPTER 9
Logarithms

9.1 Introduction to Logarithms

1 In this section, we define logarithms and prove some properties of logarithms.

2 **Definition of logarithms.** We know that $2^1 = 2$ and $2^2 = 4$. Is there a real number w such that $2^w = 3$? Such a real number exists and it is unique. We shall denote this special number by $\log_2 3$. In general, if x is a positive real number then there exists some unique real number w such that

$$2^w = x$$

We shall denote this number (w) by $\log_2 x$. Thus,

$$w = \log_2 x.$$

3 If $2^w = 2^z$ then $w = z$ for all real numbers w, z.

Proof Let $x = 2^w$. Since $2^w = 2^z$, it follows that $x = 2^z$. Now $\log_2 x = w$ and $\log_2 x = z$. Thus $w = z$. ∎

4 We now consider bases that may not be 2. If b, x are positive real numbers with $b \neq 1$ then there exists some unique real number w such that

$$b^w = x$$

We shall denote such a number by $\log_b x$.

5 If $b^w = b^z$ then $w = z$ for all real numbers w, z.

Proof Let $x = b^w$. Since $b^w = b^z$, it follows that $x = b^z$. Now $\log_b x = w$ and $\log_b x = z$. Thus $w = z$. ∎

6 For all positive real numbers b, x, with $b \neq 1$

- $\log_2 x$ is a real number w such that $2^w = x$
- $\log_3 x$ is a real number w such that $3^w = x$
- $\log_b x$ is a real number w such that $b^w = x$

7 **Properties of logarithms.** Here, we prove the following properties of logarithms:

- $\log_b 1 = 0$
- $\log_b b^w = w$

8 Let b, w be real numbers with $b > 0$ and $b \neq 1$.

9 $\log_2 1 = 0$

Proof We know that $2^0 = 1$. Using logarithm notation we may write $\log_2 1 = 0$. ■

10 $\log_b 1 = 0$

Proof We know that $b^0 = 1$. Using logarithm notation we may write $\log_b 1 = 0$. ■

11 $\log_2 2^w = w$

Proof Let

$$y = 2^w \qquad (1)$$

Using logarithm notation we may write

$$\log_2 y = w$$

Use (1) to substitute for y

$$\log_2 2^w = w$$

■

12 $\log_b b^w = w$

Proof Let

$$y = b^w \quad (1)$$

Using logarithm notation we may write

$$\log_b y = w$$

Use (1) to substitute for y

$$\log_b b^w = w$$

∎

9.2 Properties of Logarithms

1 In this section, we prove the following properties of logarithms.

- $b^{\log_b x} = x$
- $x^{\log_b y} = y^{\log_b x}$
- $\log_b y = 0$ if and only if $y = 1$.
- $\log_b x = \log_b y$ if and only if $x = y$.

2 Let b, x, y be positive real numbers with $b \neq 1$.

3 $2^{\log_2 x} = x$

Proof I

Let w be a real number such that

$$2^w = x \quad (1)$$

Thus

$$w = \log_2 x$$

Use this equation to substitute for w in (1)

$$2^{\log_2 x} = x$$

∎

Proof II By definition, $\log_2 x$ is the real number w such that $2^w = x$. Thus $2^{\log_2 x} = x$. ∎

4 $b^{\log_b x} = x$

Proof I

Let w be a real number such that

$$b^w = x \qquad (1)$$

Thus

$$w = \log_b x$$

Use this equation to substitute for w in (1)

$$b^{\log_b x} = x$$

∎

Proof II By definition, $\log_b x$ is the real number w such that $b^w = x$. Thus $b^{\log_b x} = x$. ∎

5 $x^{\log_b y} = y^{\log_b x}$

Proof

Let w, z be real numbers such that

$$b^w = x \qquad (1)$$
$$b^z = y \qquad (2)$$

Thus

$$\log_b x = w \qquad (3)$$
$$\log_b y = z \qquad (4)$$

Raise both sides of (1) to the power of z

$$(b^w)^z = x^z$$

From (4) we can substitute for z in the RHS

$$b^{wz} = x^{\log_b y} \tag{5}$$

Raise both sides of (2) to the power of w

$$(b^z)^w = y^w$$

From (3) we can substitute for w in the RHS

$$b^{wz} = y^{\log_b x}$$

From (5) and the above equation we can write

$$x^{\log_b y} = y^{\log_b x}$$

∎

6 $\log_2 y = 0$ if and only if $y = 1$.

Proof

Given $\log_2 y = 0$, then $2^0 = y$. Thus $y = 1$. Conversely, given $y = 1$ then $y = 2^0$. Hence $\log_2 y = 0$. ∎

7 $\log_b y = 0$ if and only if $y = 1$.

Proof

Given $\log_b y = 0$, then $b^0 = y$. Thus $y = 1$. Conversely, given $y = 1$ then $y = b^0$. Hence $\log_b y = 0$. ∎

8 $\log_b x = \log_b y$ if and only if $x = y$.

Proof

Let w, z be real numbers such that

$$b^w = x \tag{1}$$
$$b^z = y \tag{2}$$

Thus

$$\log_b x = w \tag{3}$$

$$\log_b y = z \tag{4}$$

We are given

$$\log_b x = \log_b y$$

Thus from (3) and (4) we obtain

$$w = z$$

Hence

$$b^w = b^z$$

From (1) and (2) we get

$$x = y$$

Conversely, if $x = y$ then $\log_b x = \log_b y$. ∎

9.3 Logarithms and Exponents

1 In this section, we prove the following properties of logarithms.

- $\log_b x^w = w \log_b x$
- $\log_b x = \log_{b^w} x^w$

2 Let b, w, x be real numbers with $b, x > 0$ and $b \neq 1$.

3 $\log_2 x^w = w \log_2 x$

Proof Let

$$z = w \log_2 x \tag{1}$$

Since $w \neq 0$ we may divide both sides by w

$$\frac{z}{w} = \log_2 x$$

Converting from logarithm notation to exponential notation, we obtain

$$2^{z/w} = x$$

9.3 Logarithms and Exponents

Raise both sides to the power w

$$(2^{z/w})^w = x^w$$
$$2^z = x^w$$

We may use logarithm notation to write the above as

$$\log_2 x^w = z$$

Use (1) to substitute for z

$$\log_2 x^w = w \log_2 x$$

∎

4   $\log_b x^w = w \log_b x$

Proof Let

$$z = w \log_b x \qquad (1)$$

Since $w \neq 0$ we may divide both sides by w

$$\frac{z}{w} = \log_b x$$

Converting from logarithm notation to exponential notation, we obtain

$$b^{z/w} = x$$

Raise both sides to the power w

$$(b^{z/w})^w = x^w$$
$$b^z = x^w$$

We may use logarithm notation to write the above as

$$\log_b x^w = z$$

Use (1) to substitute for z

$$\log_b x^w = w \log_b x$$

∎

5 $\log_b x = \log_{b^w} x^w$

Proof Let

$$z = \log_b x \qquad (1)$$

Thus,

$$b^z = x$$

Raise both sides to the power of w

$$(b^z)^w = x^w$$
$$(b^w)^z = x^w$$
$$\log_{b^w} x^w = z$$

Use (1) to substitute for z

$$\log_{b^w} x^w = \log_b x$$

■

9.4 Product and Quotient

1 In this section, we prove the following properties of logarithms.

- $\log_b(xy) = \log_b x + \log_b y$
- $\log_b \dfrac{x}{y} = \log_b x - \log_b y$ if $y \neq 0$

2 Let b, x, y be positive real numbers with $b \neq 1$.

3 $\log_2(xy) = \log_2 x + \log_2 y$

Proof Let

$$2^w = x \qquad (1)$$
$$2^z = y \qquad (2)$$

Hence using logarithm notation we may write

$$w = \log_2 x \qquad (3)$$
$$z = \log_2 y \qquad (4)$$

Multiply (1) and (2)

$$2^w \cdot 2^z = xy$$
$$2^{w+z} = xy$$

Hence we may write

$$\log_2(xy) = w + z$$

From (3) and (4) we may substitute for w and z

$$\log_2(xy) = \log_2 x + \log_2 y$$

∎

4 $\log_b(xy) = \log_b x + \log_b y$

Proof Let

$$b^w = x \quad (1)$$
$$b^z = y \quad (2)$$

Hence using logarithm notation we may write

$$w = \log_b x \quad (3)$$
$$z = \log_b y \quad (4)$$

Multiply (1) and (2)

$$b^w \cdot b^z = xy$$
$$b^{w+z} = xy$$

Hence we may write

$$\log_b(xy) = w + z$$

From (3) and (4) we may substitute for w and z

$$\log_b(xy) = \log_b x + \log_b y$$

∎

5 If $y \neq 0$ then $\log_2 \dfrac{x}{y} = \log_2 x - \log_2 y$

Proof Since $y \neq 0$, it follows that $1/y$ is a real number. Thus

$$\log_2(x(1/y)) = \log_2 x + \log_2(1/y)$$
$$= \log_2 x + \log_2 y^{-1}$$
$$\log_2 \frac{x}{y} = \log_2 x - \log_2 y$$

■

6 If $y \neq 0$ then

$$\log_b \frac{x}{y} = \log_b x - \log_b y$$

Proof Since $y \neq 0$, it follows that $1/y$ is a real number. Thus

$$\log_b(x(1/y)) = \log_b x + \log_b(1/y)$$
$$= \log_b x + \log_b y^{-1}$$
$$\log_b \frac{x}{y} = \log_b x - \log_b y$$

■

Exercises 9.4

Let b, x, y, z be positive real numbers, with $b \neq 1$. Let n be a positive integer and let $x_1, \ldots, x_n$ be positive real numbers. Prove the following.

1 $\log_b(xyz) = \log_b x + \log_b y + \log_b z$

2 $\log_b(x_1 \cdots x_n) = \log_b x_1 + \cdots + \log_b x_n$

3 $\log_b 1 + \log_b 2 + \cdots + \log_b n = \log_b(n!)$

9.5 Change of Base

1 In this section, we prove the following properties of logarithms.

- $\log_b c \cdot \log_c b = 1$
- $\log_b c \cdot \log_c x = \log_b x$

2 Let b, c, x be positive real numbers with $b, c \neq 1$.

3 $\log_2 b \cdot \log_b 2 = 1$

Proof Let

$$z = \log_2 b \qquad (1)$$
$$2^z = b$$

Since $b \neq 1$, it follows that $z \neq 0$. Hence we can raise both sides to the power of $1/z$

$$(2^z)^{1/z} = b^{1/z}$$
$$2 = b^{1/z}$$
$$\log_b 2 = \frac{1}{z}$$
$$z \log_b 2 = 1$$

Use (1) to substitute for z

$$\log_2 b \cdot \log_b 2 = 1$$

∎

4 $\log_b c \cdot \log_c b = 1$

Proof Let

$$z = \log_b c \qquad (1)$$
$$b^z = c$$

Since $c \neq 1$, it follows that $z \neq 0$. Hence we can raise both sides to the power of $1/z$

$$(b^z)^{1/z} = c^{1/z}$$
$$b = c^{1/z}$$
$$\log_c b = \frac{1}{z}$$
$$z \log_c b = 1$$

Use (1) to substitute for z

$$\log_b c \cdot \log_c b = 1$$

∎

5 $\log_3 2 \cdot \log_2 x = \log_3 x$

Proof Let

$$z = \log_3 2 \qquad (1)$$

and let

$$w = \log_2 x^z \qquad (2)$$

Thus

$$3^z = 2 \qquad (3)$$
$$2^w = x^z \qquad (4)$$

Use (3) to substitute for 2 in (4)

$$(3^z)^w = x^z$$
$$3^{wz} = x^z$$
$$\log_3 x^z = zw$$
$$z \log_3 x = zw$$

Divide both sides by z

$$\log_3 x = w$$

Use (2) to substitute for w

$$\log_3 x = \log_2 x^z$$
$$= z \log_2 x$$

Use (1) to substitute for z

$$= \log_3 2 \cdot \log_2 x$$

∎

6 $\log_b c \cdot \log_c x = \log_b x$

Proof Let

$$z = \log_b c \qquad (1)$$

and let
$$w = \log_c x^z \qquad (2)$$

Thus
$$b^z = c \qquad (3)$$
$$c^w = x^z \qquad (4)$$

Use (3) to substitute for c in (4)
$$(b^z)^w = x^z$$
$$b^{wz} = x^z$$
$$\log_b x^z = zw$$
$$z \log_b x = zw$$

Since $c \neq 1$, it follows that $z \neq 0$. Thus, we can divide both sides by z
$$\log_b x = w$$

Use (2) to substitute for w
$$\log_b x = \log_c x^z$$
$$= z \log_c x$$

Use (1) to substitute for z
$$= \log_b c \cdot \log_c x$$

∎

Exercises 9.5

Let b, c, d, x be positive real numbers, with $b, c, d \neq 1$. Prove the following.

1 $\log_d b \, \log_c d \, \log_b c = 1$

2 $\log_c d \, \log_b c \, \log_d x = \log_b x$

Made in United States
North Haven, CT
23 May 2024

52855068R00089